Einstein Lived Here

Cartoon by Herblock, published in the Washington Post some days after Einstein's death, on April 18, 1955. (Reproduced with permission from the Washington Post.)

Einstein Lived Here

ABRAHAM PAIS
Rockefeller University, New York
and
Niels Bohr Institute, Copenhagen

CLARENDON PRESS ◆ OXFORD
OXFORD UNIVERSITY PRESS ◆ NEW YORK
1994

Oxford University Press, Walton Street, Oxford OX2 6DP

Oxford New York Toronto
Delhi Bombay Calcutta Madras Karachi
Kuala Lumpur Singapore Hong Kong Tokyo
Nairobi Dar es Salaam Cape Town
Melbourne Auckland Madrid

and associated companies in
Berlin Ibadan

Oxford is a trade mark of Oxford University Press

Published in the United States
by Oxford University Press Inc., New York

A catalogue record for this book is available from the British Library

Library of Congress Cataloging in Publication Data
Pais, Abraham, 1918–
Einstein lived here : essays for the layman / Abraham Pais.
1. Einstein, Albert, 1879–1955. 2. Physics—History.
3. Physicists—Biography. I. Title.
QC16.E5P25 1994 530'.092—dc20 93–32095

ISBN 0–19–853994–0

Typeset by Footnote Graphics, Warminster, Wiltshire
Printed in The United States of America

To Ida, Joshua, and Lisa.
And a warm welcome to
young Zane Abraham.

'Why is it that nobody understands me
and everybody likes me?'

Albert Einstein, quoted in the *New York Times*,
12 March 1944

TO THE READER

On 16 April 1929, the British diplomat and author Sir Harold Nicolson wrote from Berlin to his wife, the author Vita Sackville-West, to tell her of his recent experiences. He reported having gone to a lecture by H.G. Wells, delivered in the *Reichstag* (Parliament building) and to a dinner in Wells's honor afterward at Berlin's famed hotel Adlon. 'Einstein presided. He looks like a child who for fun has put on a mask painted like Einstein. He is a darling. He made a little speech for Wells which I then translated. I began by saying: "I have been asked to translate Professor Einstein's speech. I may add that it is the first thing of his I have understood." They thought that a funny joke.'

There is a profusion of those who, like Sir Harold, have understood little of what Einstein has said, thought, and done, and who are hungry to know more about him. Since I have understood more about Einstein than many, not just because I am a physicist, but more so because I knew him well in his later years, I attempt in this book to bring the general public closer to Einstein the man. With this audience in mind, I have scrupulously avoided using any mathematics – with the exception of the inevitable formula $E=mc^2$ and a few equations in Chapter 6.

When, far along in the preparation of this collection of essays, my eyes fell on Einstein's query, I knew at once that I had also found the epigraph to this volume, even though this statement is not precise. It is neither true that no one understands Einstein nor that everyone loves him. It is not rare to find a solid course on relativity theory advertised among the graduate offerings by good universities. As to examples of people who disliked the old man, the attentive reader will in fact find several of these in the following pages.

This book is a companion volume to my *Subtle is the Lord* (Clarendon Press, Oxford, 1982), but not its sequel. Earlier I had focused on Einstein's science and his life and had only occasionally included remarks about the way he was perceived by the outside world of non-scientists. The main purpose of the present volume is to enlarge on that last topic, which now becomes the central theme.

There exists, however, some overlap with my previous book, from which I have borrowed two of the following essays, the one on Einstein and Newton (Chapter 4), and the one on the Nobel prize (Chapter 6). In view of the potentially different audience this time, it has seemed appropriate to me to lift these out of the earlier work. In the other essays the reader who is acquainted with *Subtle is the Lord* may find familiar phrases as well, but only rarely.

Some entries included here deal with material that was not yet available in 1982, when my earlier book on Einstein was concluded, notably that he had an illegitimate daughter by his later first wife, a fact that entered the public domain only in 1986. In that year a number of high-school classmates published a collection of reminiscences about Einstein's younger son, who had become manifestly schizophrenic just after the years of their close contact with him. Their book has so far been published only in German. Also of recent vintage are publications in which it is alleged that Einstein's first wife played an important role in his formulation of relativity theory. I deal with these three topics in the essay 'In the Shadow of Albert Einstein' (Chapter 1).

It had become evident to me in the course of writing my previous book on Einstein that the world-wide nature of his renown was the result of the attention he had received from the media. The thought had occurred to me already then that it might be quite illuminating to explore this press coverage more fully. Accordingly I have, every now and then during the past ten years, delved in newspaper and magazine archives, beginning with microfilm of the *New York Times* from 1919, the year in which Einstein made his first appearance in that newspaper of record, until the present. By the late 1980s, many hours of searching had produced some fifty sheets of tightly handwritten notes.

Meanwhile I had become convinced that I should try to extend this material as far back in Einstein's life as possible and also to make efforts at obtaining clippings from the foreign press. With wonderful help from local spies I found out that as early as 1902 Einstein had made his first appearance in a newspaper. Fascinating tidbits emerged from Swiss, Czech, and German dailies – the countries where Einstein had held academic positions prior to his coming to the United States.

All this material has led to 'Einstein and the Press', the main essay in this volume and the one that has taken the most labor to put together (Chapter 11).

Interviews with Einstein as well as reports on brief comments and longer addresses by him will, I hope, convey his vivid style of expression as well as his great talent at formulation. I also hope that my readers will share my fascination with getting to know other's opinions of the man, including but not at all confined to his science. Indeed, the press has always been delectated to report on his pronouncements regarding pacifism, supranationalism, civil liberties, and the rights and obligations of Jews and Arabs to live together in harmony and dignity in the Middle East, the main themes of Einstein's extra-scientific thinking. That is still not all, however. I frequently interrupt my account with press reports on varia dealing with his opinions on subjects ranging from capital punishment to vegetarianism. Stay tuned.

Opinions are bound to differ as to whether Einstein's pronouncements on this plethora of issues are all that important. I have aimed at conveying their totality since that is the best way, I believe, of giving the reader a feel for the workings of this most unusual mind. Whatever may long be remembered or soon forgotten, his lasting greatness will be his science.

I gratefully record the assistance of those who helped me get this book in shape. Warm thanks go to members of the Einstein Studies program at Boston University, in particular to Robert Schulmann for his most useful, friendly counsel and to Annette Pringle, who collected materials for me. Ze'ev Rosenkrantz, from the Hebrew University, was good enough to provide me with documents from Jerusalem. Dr and Mrs Karlheinz Steinmüller from Berlin helped me gather clippings from German newspapers. I also thank Ambassador Hans Henrik Bruun for drawing my attention to the Nicolson letter from which I just quoted. I was most fortunate to have Ms Jan Maier work for me on preparing the manuscript and for her helpful editorial comments in the process.

I am much beholden to the Alfred P. Sloan Foundation for a grant that helped me in many phases of preparation.

Finally, I should like to acknowledge permission to quote or reprint from material in the Albert Einstein Archives, both in Boston and in Jerusalem; the Fondation Louis de Broglie; the Cambridge University Press; and *American Scientist*. These institutions are mentioned at the beginning of the relevant essays.

Thank you, my dear Ida, for your constant support, not least for your wise questioning as the writing proceeded!

New York
February 1993

A.P.

References

Each chapter has its own set of references. Abbreviations frequently used are:

CP, *Collected papers of Albert Einstein*, ed. J. Stachel *et al.*, Princeton University Press, 1987.

NYT, *New York Times*.

RS, J. Renn and R. Schulmann, *Albert Einstein–Mileva Marić, the love letters*, Princeton University Press, 1992.

SL, *Subtle is the Lord*, A. Pais, Clarendon Press, Oxford, 1982.

VZ, *Vossische Zeitung*.

CONTENTS

ILLUSTRATIONS

Portrait of Einstein by Karsh from Ottowa, early 1950s. (Courtesy of Karsh/Camera Press.)

1

'In the shadow of Albert Einstein'

1. Introduction

In 1969 the publishing house Bagdala in Kruševac (Yugoslavia) brought out a biography of Mileva Marić, Albert Einstein's first wife. The book is written in Serbo-Croatian.* In English translation, its title is *In the shadow of Albert Einstein*. Its author was a Serbian woman, Desanka Trbuhović-Gjurić, a retired middle-school teacher of mathematics. Some years later the typescript of a German translation began to circulate. I read this document prior to finishing my Einstein biography,[1] but was not greatly impressed by its content and did not mention it in my book.

Meanwhile a German translation of this Marić biography has appeared in print. Of particular interest is its recent fourth edition,[2] which came out in 1988, after the death of the author, and after the appearance of the first volume of Einstein's collected works.[3] This last book contains new intelligence about the relations between Albert and Mileva, none of which was in the public domain during Trbuhović's life. Of crucial importance in this book are the 41 letters from Albert to Mileva, and the 10 from her to him, exchanged in the period that ends in 1902, after the birth out of wedlock of their daughter Lieserl. These and other documents were uncovered in 1986 in a bank vault in San Francisco, among papers of the estate of Einstein's elder son Hans Albert.

This new information has for the first time been incorporated in the fourth German edition of the Mileva biography, by Werner Zimmermann, archivist (since retired) of the city of Zürich, and titular professor of Slavic languages and literature at the University there. He has written addenda to several chapters and an overall epilog, thereby greatly improving the quality and perspective of this 200-page volume.

The biography of Mileva gives interesting and unique information about

* The original title is *U senci Alberta Ajnštajna*.

Earliest known photograph of Einstein, early 1880s. (Bilderdienst Süddeutscher Verlag.)

her family background and early years. The final third, devoted to her difficult late years, is likewise commendable. Unfortunately, the same cannot be said about the book's central theme: the nature of the relationship between Einstein and the woman who was to be his first wife. What is written about the role of Mileva in regard to Albert's scientific output is surprising and astonishing. I have not found a single reason for believing that in this respect the author's allegations are founded on fact. That is why, after reading the early typescript, I decided it was wisest not to refer to what after all appeared to be a draft, hoping, vainly as it turned out, that a final version would be toned down.

There are several reasons why now, a decade after having finished my earlier book,[1] I nevertheless come back to the Mileva biography.

First, it gives me an opportunity to comment on new material not available to me before.

Secondly, in recent years a minor industry has evolved aimed at demonstrating the important influence of Mileva on Einstein's work, particularly on his relativity theory. All these recent claims, based in the first instance on what Trbuhović started, *were never made by Mileva herself*. I have no interest in entering into polemics on these matters, since I am convinced that this outpouring will eventually fade away without leaving a trace. Nevertheless I think it may serve some purpose to spell out what the actual facts are as I see them.

Thirdly, it has come to my notice that the book on Mileva leaves to be desired in another respect, to wit, the account of Albert's and Mileva's younger son, Eduard, another tragic figure. I became aware of this from reading a monograph,[4] published in 1986, written by nine of his high-school classmates, all in their mid-seventies when their book came out. In the preface to their joint venture, these men note, courteously but firmly, that Trbuhović's sources are inadequate. In this essay I shall also report on these biographical sketches of Eduard, which are not well known in the English-speaking world.

It may be helpful to begin by recalling some chronology.

2. Some background

Albert Einstein was born on Friday, 14 March 1879, in his parental home, Bahnhofstrasse B 135, in the city of Ulm in the German state of Württemberg. His father, Hermann, was a small businessman, never very successful. His mother was Pauline née Koch. In 1880 the family moved to Munich, where, on 18 November 1881, Maria (always called Maja), his only sibling, was born. Of all the women who were to play a role in Einstein's life, Maja was the one to whom he always felt closest. In general terms one may characterize his childhood milieu as warm, stable, and stimulating.

At the age of five Einstein received his first instruction, at home from a woman teacher. At the age of seven he entered the *Volksschule*, a public school, where he did very well. In 1886 Einstein's mother wrote in a letter that her son was 'again number one'.[5] In 1888 he went on to high school, at the Luitpold *Gymnasium*, where again he did well. When in 1894 his family moved to Milan, they left Albert in the care of a Munich family, so that he could continue his schooling. In a biographical sketch[6] of her brother, completed in 1924, Maja has recalled that this solitary existence depressed Einstein and made him nervous. So half a year later he withdrew from his school, at his own initiative, and joined his family in Italy. He was determined, however, to continue preparations for admission to a university. This he did by self-study.[7]

In October 1895 Einstein took by special permission – he was two years under the minimum age of admission – the entrance examination for the *Polytechnikum* in Zürich, in 1911 renamed the *Eidgenössische Technische Hochschule* (ETH, the Federal Institute of Technology). He was examined on scientific subjects (mathematics, physics, and chemistry) and on general subjects (literary and political history, German). He failed because of inadequate performance in the latter topics. This setback may perhaps have caused the rather widespread erroneous impression that Einstein was a poor school student. It was not that he ever did poorly – it was rather that he never liked school.

Following the advice of the Director of the ETH, Einstein next spent one year at a school in Aarau, in Switzerland, to finish his secondary education. In September 1896 he passed the final school examination, which entitled him to university admission. With the exception of French, his grades were fine in all other topics, especially mathematics, physics, singing, and music (violin).[8]

In October Einstein now finally enrolled in Section VIA of the ETH, which provided training for a degree as a high-school teacher in physics and mathematics.[9] He was the youngest of a group of five to enter that Section, still only seventeen. The oldest was Mileva.*

Mileva was born on 19 December 1875, in the town of Titel, in the Vojvodina, now a region of Yugoslavia, then in Hungary, which at the time was part of the Austro-Hungarian empire. She was the eldest child of Serbian parents. Milos Marić, her father, was a middle-level administrative government official. Her mother, Marija, née Ruzić, was the daughter of a well-to-do land owner.

Mileva was born with a luxated hip joint which at that time could not be medically corrected. She limped all her life. So did her younger sister Zorka, who never married. Late in life Zorka became severely mentally ill, and died surrounded only by her many cats.

The sisters had one younger brother, Miloš. Apart from his native Serbian, he learned to speak Hungarian, German, French, and English. He went on to study medicine, partly in France, where in 1914 he married a frenchwoman. When the First World War broke out he had to join the army in 1914, where he became medical doctor attached to a battalion. That same year he was taken prisoner of war by the Russians. In 1915 he was allowed to work in Russian hospitals. After the war he continued a medical career in the Soviet Union, eventually becoming professor of

* The other three were Marcel Grossmann and Louis Kollros, who both became ETH professors, and Jakob Ehrat, who became a mathematics teacher.

Mileva Marić in 1899. (Schweizerische Landesbibliothek.)

histology at the medical institute in Saratov on the Volga, 500 miles south-east of Moscow.[10] He never returned to his native land and by postcard informed his wife that she should consider herself as free of him. The Marićs were not a run-of-the-mill family.

Mileva's schooling started in 1882. Classes were given in Serbo-Croatian but she also learned German and French. It appears that she was a good and enterprising student. In order to round off her high-school education, her father sent her to Zürich, where in November 1894 she was admitted to the *Höhere Töchterschule* (advanced school for girls). In the spring of 1896 she passed the final examination entitling her to enter university. That summer she enrolled as a medical student in the University of Zürich, an institution that has the distinction of being the first in Europe to produce a woman Ph.D. (in 1867). At least two Serbian women had obtained that degree there even before Mileva entered. In the autumn of 1896 she had already changed her study plans from medicine to mathematics and physics. To this end she switched to section VIA at the ETH – where she met Einstein.[11]

Leaving until later the consequences of this encounter, I relate next what little I know of Mileva's studies at the ETH. The register of that institution lists her marks in various courses for each of the academic years 1896–7 to 1899–1900. They are quite good, comparable to those of Einstein.[12] From October 1897 to February 1898 she studied at the University of Heidelberg, following courses in mathematics and physics.

In July 1900 the classmates took their final examination. Albert passed, Mileva failed. Her marks were decent with the exception of mathematics.[13] In July 1901 she tried again (under very trying conditions; see section (4) of this chapter). Once more she did not succeed. After this she gave up trying. Nor did thereafter her name appear, alone or as co-author, on any scientific publication.

3. Mileva and Albert, from first acquaintance to the arrival of Lieserl

The first intimation that the relation between the two was beginning to exceed those of fellow-students is found in a letter[14]* written in the autumn of 1897 by Mileva, then in Heidelberg, to Einstein, in which she told him that she had spoken to her father about him and that he should 'by all means come along' for a visit. Einstein's first preserved letter to her[15] dates from February 1898. Its salutation: *'Geehrtes Fräulein'* (literally: 'honored Miss') is formal. He wrote of the ETH courses and asked for a speedy reply. In his next letter[16] he addressed her as 'Dear Miss Marić,' and invited her to dinner in his boarding-house. In March 1899 he wrote again, this time from Milan, where he was on a family visit: 'Your picture has made a great impression on my parents . . . I had to endure some teasing which did not displease me at all.' Earlier he had signed himself 'Albert Einstein', this time: 'Albert'.[17]

In August 1899, when each of them was on a family holiday, he wrote for the first time about reading physics texts together and also about family relations. 'I find my mother and sister a bit narrow-minded . . . It is remarkable how family ties weaken and how one's inner life becomes so mutually incomprehensible that one can in no way feel what moves the other.'[18]

Another letter,[19] one week later, is interesting for several reasons. Einstein writes it to 'L.D.', which stands for *Liebes Doxerl*, 'Dear little doll', an expression used in Swabian – in which *Docke* means 'doll' – the dialect of his native Württemberg, Bavaria, and of North-west Tyrol. This form of language must have been dear to him; all through his life he spoke German with a marked Swabian accent.

Particularly fascinating is the presentiment of a basic trait of Einstein's

* This and later letters have meanwhile appeared in English translation.[3]

special relativity theory which he was to set forth in 1905: 'The introduction of the name aether in electrical theories has led to the conception of a medium which can be considered to be capable of motion. I believe one cannot give physical meaning to this statement.'[19]

Here, just before the close of the nineteenth century, we see the beginning of doubts about the aether that would eventually lead Einstein to abolish this medium altogether.

Mileva's reply[20] to this letter begins more formally with 'LHE', '*Lieber Herr* E.', 'Dear Mr. E.' Further on we read, however: 'From common experience a still quite furtive curious feeling has developed . . .' This letter further contains greetings to Albert's mother and sister, but no comments on physics. Nor did she react to remarks in his next two letters concerning 'a good idea for the investigation of the motion of bodies relative to the light aether.'[21]

The correspondence shows that toward the end of 1899 their relationship had warmed up considerably. Einstein greets Mileva as 'My sweet *Doxerl*', and writes of 'our housekeeping'.[22] At some time, probably in 1900, Mileva sent a little note[23] to Albert beginning with '*Mei liebs Johonesl*', 'My dear J.', and ending: '*Tausend Küserlein von deins D*' ('A thousand little kisses from your D'). Note that she had dropped the formal *Sie* for the familiar *Du*. *Johonesl* is a diminutive for Johannes. During the next year their writing style becomes increasingly cloying. Their letters are filled with diminutives and pet names like *Miezchen*, little cat, for Mileva. While this was going on, young Albert was not averse from a little flirtation on the side.[24]

The next letters from Einstein date from late July and August 1900, just after Mileva had failed her examination, a fact which, curiously enough, he does not mention at all. The main theme is now his parents' reaction to his private life. After he had arrived at the hotel where his mother was on holiday, she asked what was happening with *Doxerl*. '"My wife", I say innocently but prepared for a scene. Mama threw herself on the bed and cried like a baby. When she had recovered from the first shock, she immediately went on a desperate offensive: "You are ruining your future and blocking your life's path . . . she does not fit in a decent family . . . you will be in a fine mess when she gets a child . . ." I energetically rejected the suspicion that we had lived together indecently.'[25] In a following letter he reported that he had received a moralizing letter from his father. 'I understand my parents very well. They consider a woman as a man's luxury which he can only afford after having found a comfortable position.' He went on to urge Mileva not to talk to her parents, so as to avoid similar complications, and added, 'When I do not have you I feel as if I do not quite exist.'[26] A few days later he told Mileva that his mother now maintained complete silence on the issue, and that he missed 'my little dear "right hand"'.[27]

As mentioned above, in July 1900 Einstein had obtained his diploma as science teacher. The time had therefore come for him to find a job. That undertaking turned out to cause him quite a bit of trouble. Efforts to procure an assistantship at various universities all failed. Not until May 1901 did he have his first position, and that only a temporary one, as substitute high-school teacher in the Swiss town of Winterthur. The following October he started teaching at a private school in Schaffhausen, where he stayed until January 1902. Only in June 1902 did he acquire a more permanent position, a junior appointment at the Patent Office in Berne.[28]

Meanwhile he had begun his own independent research. It caused him satisfaction amidst all this confusion when in December 1900 he could submit his first article to a scientific journal: a not too important paper on the theory of capillarity.

In those years he had occasion to write to Mileva not only when he was on holiday with his family but also while he was holding down his temporary jobs away from Zürich, where Mileva continued to reside in that period. Together with the themes already mentioned, new ones now appear in his letters: concerns about finding employment, enthusiasms about physics research.

It is worthy of note how in those years he draws Mileva in, as it were, into his own research projects: 'Whatever will happen, we shall have the most delightful life in the world. Good work and togetherness.'[29] About his first paper: 'When it is ready *we* [my italics] send it in to *Wiedemann's Annalen*'[30]* . . . 'Our paper . . . if only we shall soon have the good fortune to move forward together along this beautiful path.'[31] About a project for a subsequent paper: 'Our theory of molecular forces . . .'[32]

Particularly interesting is a reference in an early letter by Einstein (from 1901) about collaboration on an issue that would lead to his 1905 theory of relativity: 'How happy and proud I shall be when together we shall conclude victoriously our work on relative motion.'[33] The earliest mention I have seen of his touching on relativity is in this letter and in an even earlier one, also to Mileva (of 28 September 1899).[21]

Another theme that keeps recurring is the attitudes of Albert's parents. 'My parents weep for me almost as if I had died'[34] . . . 'My parents have accepted the inevitable . . . I am glad I have told everything.'[35] Mileva to Albert: 'I do not believe that your mother will reconcile herself with you.'[36] Some months thereafter she wrote to a friend about mama Einstein:

* An earlier name for the prestigious *Annalen der Physik*.

'This lady seems to have made it her life's purpose to embitter the life not only of me but also of her son . . . They managed to write a letter to my parents in which they revile me to a scandalous extent.'[37] Nor were Mileva's own parents all that happy. Einstein to Mileva, November 1901: 'I am very glad that your parents have calmed down somewhat and now have more confidence in me. I know I deserve that.'[38]

Otherwise Einstein's letters of 1900–1 express his love: 'How could I have lived before'[39] . . . 'Without thoughts of you I would rather not live any more'[40] . . . 'Among everybody you love me most and understand me best'[33] . . . 'Evenings I think that [you] think of me with love and in bed kiss his pillow. I know how that goes!'[41] . . . 'My happiness is your happiness'[42] . . . 'My life gets its true content only because of thoughts about you . . . How wonderful it was last time when I could hold you as Nature has given it (*Wie die Natur es gegeben hat*).'[43] Mileva to Albert: 'How wonderful the world will look when I shall be your little wife'[44] . . . 'I shall invite you to see my family during the holidays. Wouldn't you like to come along a bit?'[45]

4. Lieserl

In May 1901 a new theme enters the correspondence which, it is safe to guess, came totally unanticipated to the young people.

In a letter[46] which, typically, begins with the expressions of enthusiasm about a paper on cathode rays Einstein had just read, he continues: 'How is the boy? . . . How is it going with our little son and your paper for the Ph.D.?'* A week later: 'How is it with your studies and with your little child?'[47]

Mileva was pregnant.

Never in Einstein's life must the future have looked more uncertain to him than in that spring of 1901. All that his efforts at finding work had produced so far was a temporary position as high-school teacher in Winterthur. Now he was soon to be a father . . .

A month later he wrote: 'I have decided the following about our future. I shall look for a position, however poor . . . As soon as I have found one I marry you . . . Although our situation is very difficult I am quite confident since having made that decision.'[48]

Shortly afterward he wrote to Mileva to wish her all the best with her forthcoming examination.[49] This referred to her second attempt at obtain-

* ETH graduates could obtain a Ph.D. from the University of Zürich without further examination. All they were required to do was to submit a Ph.D. thesis for approval. Einstein's letter[47] was written shortly before Mileva made her second attempt to graduate from the ETH.

ing a teacher's diploma. As mentioned above, she failed again. Now that we know that at that time she was in an interesting condition we must admire all the more her courage and persistence in trying anew.

The letter just quoted[47] shows that Einstein was hoping for a son. Not so Mileva. In December 1901, now teaching in Schaffhausen, Albert wrote: 'I look forward to our dear Lieserl which I, however, secretly (so that *Doxerl* does not know it) imagine as a Hanserl . . . All that remains to be resolved is the question how we can take in our Lieserl; I would not like to give her away . . . We remain students as long as we live and don't give a shit (*kümmern uns keinen Dreck*) for the world.'[50] A week later: 'I would so much like to have you with me in spite of your "funny figure" . . . Draw it for me! . . . I am working on the electrodynamics of moving bodies.'[51]

That week Einstein received good news: a position for him at the Patent Office in Berne seemed assured. 'Only now do I realize how much I love you, since the strong pressure of the situation no longer burdens me.'[52]

A letter from the closing days of 1901 shows how strongly Einstein's existence is focused on Mileva: 'All people except you appear strange to me, as if they were separated from me by an invisible wall.'[53]

Shortly thereafter he moved from Schaffhausen to Berne. There, in January 1902, word reached him that he had become the father of a daughter. She was named Lieserl, evidently Mileva's choice.

Mileva had gone to her home region in Serbia to give birth.[54] Einstein wrote to her there. 'It has really become a Lieserl. Is she healthy and cries properly? Who gives her milk? Is it hungry? I love her so and do not know her yet . . . I should like to make a Lieserl myself, that must be interesting! It certainly can cry already but will learn to laugh only much later. Therein lies a deep truth.'[54] We also learn from this letter that Mileva was not well after having given birth.

I am not sure when or how the news reached Einstein's parents. Perhaps his mother knew already in November–December 1901 what was about to happen when she had written to Mileva's parents.[37] Perhaps his mother knew in late February, when she wrote to a friend: 'That miss Marić gives me the bitterest hours of my life, if it were in my power I would do all I could to ban her from our horizon.'[55]

The Einstein–Marić letters printed in the much-quoted Volume 1 of the collected papers are but a small part of their correspondence that was unearthed in 1986. More of the other letters found (some 400), to Mileva and to their sons, will be published in subsequent volumes. The discovery of these letters delayed the printing of Volume 1. Since that volume deals with the period of Einstein's life that ends in 1902 it was clearly essential to incorporate the appropriate fraction of this material. This was one of the reasons for extra delay in the publication of the volume.

Another was the fact that the very existence of Lieserl came as a total surprise to all Einstein experts. The obvious question arose at once: What became of her? Can one publish Volume 1 without knowing that?

Members of the editorial staff immediately set wheels in motion to find out. This included trips to Yugoslavia and Hungary. All efforts to catch even a glimpse of the girl's later life were in vain, however, and so it has remained to this day.

When it became clear that one *had* to proceed with the publication of Volume 1 without this further information, the delicate question arose of how to handle the newspaper press which, one way or another, would get wind of the existence at one time of the little daughter whose later fate was and is shrouded in mystery. During discussions by the Editorial Committee of the Einstein papers, of which I am a member, I suggested that it would be best to inform the press straight away since the papers would get hold of the story in any case. I saw little harm in their publishing this information. There would, I expected, be a very brief flare-up, after which the whole matter would rapidly be forgotten. And that is indeed what happened. The *New York Times* of Sunday, 3 May 1987 carried a front page article under the heading: 'Einstein Letters Tell of Anguished Love Affair', gave a sensible account of what was known; and that, more or less, was that.

5. Marriage. Hans Albert. What became of Lieserl?

On 10 October 1902, Einstein's father died, felled by a brief heart disease. When the end was near he asked everyone to leave so that he could die alone. It was a moment Albert never recalled without feelings of guilt.[56]

On 6 January 1903, Albert and Mileva were married in Bern, in a civil ceremony. They sent out wedding announcements printed in old-fashioned style.[57] Their witnesses* were Conrad Habicht and Maurice Solovine, two friends in common. There was a small party that evening. Afterward, when the couple arrived at their lodgings in Archivstrasse 8, Einstein had to wake up the landlord. He had forgotten his keys . . . .[58] Late in life Einstein stated that he had entered the marriage with inner resistance.[59]

In 1906 Einstein's sister Maja also moved to Berne, in order to continue the university studies she had begun two years earlier in Berlin. She was clearly a gifted woman. On 21 December 1908, she received her Ph.D. *magna cum laude* on a thesis in Romance languages, which was subsequently

* As is seen on the marriage protocol.[58]

Wedding photograph of Einstein and Mileva Marić, January 6, 1903. (©Evelyn Einstein.)

published by a German firm.[60] In 1910 she married Paul Winteler, a Swiss. Their union remained childless.

Meanwhile, on 14 May 1904, Albert and Mileva's son Hans Albert had been born. He was to become a distinguished scientist in his own right. In 1926 he obtained an ETH diploma as civil engineer. In 1928 he married Frida Knecht in Dortmund, where he worked for some time as a designer of steel works. In 1930 their son Bernhard Caesar was born, and through him the family line continues to this day. (He currently lives near Berne.) In 1936 Hans Albert received a Ph.D. from the ETH. In 1938 he and his family emigrated to the United States, where his wife died. Soon after her death he married Elizabeth Roboz.[60a] From 1947–71 he was professor of hydraulic engineering at the University of California in Berkeley. In 1973 he died of a heart attack and was buried on Martha's Vineyard. On his graveyard one finds the phrase: 'A life devoted to his students, research, nature, and music'.

About his father's influence Hans Albert once remarked, 'Probably the only project he ever gave up on was me. He tried to give me advice, but he soon discovered that I was too stubborn and that he was just wasting his time.'[61]

The relations between father and son were not always good. I have seen a rather nasty letter[62] (written long after his parents had divorced, an event to which I shall come back) in which Hans Albert blamed Einstein for not

Einstein with his first-born son, Hans Albert, late 1904 or early 1905. (©Evelyn Einstein.)

sufficiently securing the material future of Mileva and his children.

On Saturday, 16 April 1955, Hans Albert arrived in Princeton, where his father lay dying in the hospital. 'On Saturday and Sunday I was together quite a lot with my father who much enjoyed my company.'[63] Einstein died in the early morning hours of 18 April . . .

What became of Lieserl?

Nothing I have seen indicates that Einstein ever even saw her.

After their marriage, the Einsteins did not have her with them in Bern.

In the summer of 1903 Mileva went on a visit to her family. During that period Einstein wrote to her[64] from Bern expressing concern about Lieserl's attack of scarlet fever. He asked how the little girl had been registered, and told Mileva that he would see to it that she would get a new Lieserl who should not be denied her. (At the time Einstein wrote, Mileva was in fact already pregnant with Hans Albert!)

This is the last known communication between the parents about their little girl.

Einstein with his stepdaughter Margot and his son Hans Albert, during the latter's 1937 visit to the U.S. Picture taken on the front porch at Mercer Street, Princeton. (Einstein Archive, Boston.)

6. Of Albert, Mileva, and the theory of relativity

In the summer of 1903 the Einsteins moved into an apartment at Kramgasse 49, which is still the most beautiful old street in Berne. It was there that in 1905 Einstein did his work on the special theory of relativity. I now turn to a discussion of those days.

Earlier I quoted from a letter[19] by Einstein to Mileva written in August 1899 in which for the first time he expressed doubts about the physical reality of the aether. Compare this with Trbuhović's statement: 'It is she who raises the question about the aether . . . Albert picks up this problem and together they seek for its solution.'[65] There is no documentary evidence, however, that it was Mileva who initiated this investigation. According to Trbuhović, Einstein had once said to Mileva's brother Miloš: 'She was the first to draw my attention to the significance of the aether in the universe'[66] – but no source for this oral statement is given. Trbuhović again: 'All that Albert created later originated from what was achieved with Mileva's immediate collaboration and which later was developed in longer periods of time . . . In his work she was not the co-creator of his ideas . . . but checked them all . . . and gave mathematical expression to his concepts of the special theory of relativity.'[67]

I simply cannot believe that. Einstein himself was perfectly capable of handling the mathematics he needed. He said to me once, much later, 'I am not a mathematician.' Yet when the need arose he could master more than sufficiently the mathematical tools, as is particularly seen in his early work on the general theory of relativity. Furthermore, it is part of the charm of special relativity that its mathematics is quite elementary; the subtlety lies in the concepts. Moreover, is it plausible to expect Mileva to be his helper with the mathematics, in view of the fact that this was the very subject in which her inadequate marks caused her to fail in obtaining her ETH diploma?

Trbuhović's most provocative allegation concerning Mileva's role is her claim about a statement supposed to have appeared in the memoirs by the distinguished Russian physicist Abraham Joffe: 'Joffe notes that the three epochal articles by Einstein from 1905 were signed "Einstein–Marić." Joffe had seen the originals when he was assistant to Röntgen who was member of the editorial board of the *Annalen [der Physik]*.'[68] If true that would be a piece of evidence which would certainly give food for reflection. I therefore wrote to my friend Robert Schulmann, co-editor of the Einstein volumes, to ask his opinion. Schulmann replied: 'The passage in the book which interests you also exercised me on first reading it and when I interviewed the lady in 1983 I asked about this passage. She gave me what I can only call charitably an extremely weak explanation. She said that the proof of the statement was contained in a microfilm which she had had to return to the Soviet Union a few weeks before.'[69] Later I had occasion to raise the same question with Professor Victor Frankel, the well-known Russian historian of science. He had searched for this reference to Mileva in the Joffe memoirs and had not found anything.

Nevertheless Joffe did write something pertinent which, however, Trbuhović interpreted incorrectly (if that is what she saw). Shortly after Einstein's death he published a short memorial[70]* in which he stated, 'In 1905 there appeared three articles in the *Annalen der Physik* . . . the author

* I owe this important reference to Robert Schulmann.

of these articles – unknown at that time – was the clerk at the Patent Bureau Einstein–Marity',* and adds, in parentheses: '(*Marity – the family name of his wife, which by Swiss custom is added to the husband's family name*)' [my italics]. This is clearly a reference to a single author. Alas, I do not know what caused Joffe to refer to 'Einstein–Marity'.

All that remains, then, as evidence for a possible role of Mileva in the creation of relativity is Einstein's remark, quoted before, in a letter [33] of March 1901: 'Together we shall conclude victoriously our work on relative motion.' In judging the weight of this comment one should remember, first, that at that time Einstein was a young man very much in love; secondly that at that time of great uncertainty in his life he had hardly any contact with scientists of his generation. Would it not suffice to deduce that Mileva was the most welcome figure in his life at that time, not so much as collaborator but as lover as well as the only readily available sounding-board for his ideas?

Why then should Trbuhović go to such length to stress Mileva's role? I think she herself has given the answer: 'We cannot avoid to take pride in our great Serbian Mileva Marić who participated in the creation and formulation [of Einstein's work].'[71] It is a sentiment one can understand even if one cannot condone it.

7. Separation. Divorce. Einstein remarries

First I sketch in a few lines Einstein's career during the following years.

In 1908 he was admitted as *Privatdozent* (a non-faculty teaching position) at the University of Berne. In October 1909 he and his family moved to Zürich, where he had been appointed to associate professor in the University. It is the city of which Mileva grew quite fond and in which she was to spend nearly all the rest of her life. This was the time when Einstein's fame began to spread. In that year he received his first honorary doctorate (from the University of Geneva). The next year he was for the first time nominated for a Nobel prize and, on 28 July, his second son was born. He was named Eduard; more about him later on.

In Zürich the Einsteins had a stimulating social life. In particular they often visited the home of Adolf Hurwitz, who had been one of the Einsteins' mathematics professors at the ETH. He and Einstein shared not only scientific interests but also a love of music. Einstein would often bring along his violin to participate in playing chamber music at the Hurwitz home. One of the most reliable sources on Einstein's life in that period,

* This Hungarian spelling of the Serbian *Marić* is found earlier in the marriage protocol.[58]

frequently quoted by Trbuhović, are the diaries kept by Hurwitz's elder daughter Lisbeth, who met Einstein for the first time in January 1911, and who took dancing lessons in a class also attended by Hans Albert.[72] 'Albert Einstein comes regularly [once a week] and we play Bach and Mozart.'[73]

In March 1911 the Einsteins were once again on the move, this time to Prague, where Einstein had been appointed full professor in the German University. They stayed there only for about one-and-a-half years, after which they returned to Zürich, where Einstein had obtained a full professorship, this time at the ETH. In those years Mileva occasionally accompanied her husband on journeys abroad for his scientific purposes, such as a trip to Leiden for a visit to the grand old master Hendrik Antoon Lorentz.[74]

As far as I know the marital relations between the Einsteins were no cause for serious concern in those early years. A change appears to have begun after the move to Prague. According to an Einstein biographer: 'Various mines laid by outsiders to spring the marriage now began to work.'[75] Especially because Einstein was now more and more becoming a public figure, his family did not think highly of the spouse at his side, who not only limped but also began to neglect her appearance, a trend that would worsen as she grew older. It is also known that Mileva had left Zürich only with reluctance. As Trbuhović plausibly conjectures,[76] in Prague she was torn between the preponderantly German milieu of her husband and the dominant Slav environment closer to her own roots. Germans constituted only 20 per cent of the Prague population; in the 1880s tensions had caused the University to split into a Czech and a German institution with little interaction.

We shall see shortly that there was a more important reason for the estrangement between Einstein and his wife while they were in Prague.

As mentioned, the brief Prague interlude was followed by the return to Zürich. Old ties were reknit, music hours at the Hurwitzes were re-established. Mileva was not well now. She walked with difficulty because of severe leg pains, and had become melancholic.[77] In March 1913 she wrote to a friend that her husband now lived only for his science and paid little attention to his family.[78]

On 16 March 1914, Lisbeth Hurwitz noted in her diary[79] that Einstein had come one last time to make music. He was about to leave for Berlin where he had accepted a major academic appointment under highly favorable conditions.[80] Later that month he moved to the German capital, where he was soon joined by his wife and their two sons.

Whatever hopes Mileva may have had for better relations with her husband resulting from the return to Zürich, the one-and-a-half years spent there had not improved the situation, which worsened to a crisis after the move to Berlin. From that time dates a draft of what one can only call a memorandum by Einstein to Mileva in which conditions for continued coexistence are set forth. She is to renounce personal relations with him unless social obligations demand otherwise. She should not expect any tenderness from his side; there would be no more joint going out or travel: rough stuff.

The situation had manifestly become impossible. Separation followed. In June 1914 Mileva and the two boys left Berlin to return to Zürich, which remained her home for the rest of her life.

Einstein accompanied Mileva and his two sons to the Berlin railway station. Helen Dukas, from 1928 on Einstein's secretary and later also his housekeeper, once told me that '*weinend ist er vom Bahnhof zurückgegangen*' (he cried on the way back from the station). The departure of Mileva must have come as a great relief to him, but the separation from his children caused him much sorrow.

By the time Mileva left, Einstein's future second wife was already very much on the scene.

Elsa Einstein was born in 1876 in Hechingen, in Hohenzollern. She was both first and second cousin of Albert. Rudolf, her father, a well-to-do director of a textile factory in Hechingen, was first cousin of Albert's father. Fanny, her mother, was a sister of his mother Pauline. Elsa and Albert had known and been fond of each other since childhood. She had married Max Löwenthal, a merchant, by whom she had two daughters, Ilse and Margot. Most probably Elsa did not receive any higher education, since at the age of 21 she had already entered the blessed state of motherhood. After her brief marriage had ended in divorce in 1908, she and her two daughters had readopted the name Einstein.

Elsa and her daughters were living in the Haberlandstrasse in Berlin at the time Einstein and his family arrived there. By then he had been in steady contact with Elsa for the preceding two years.

In 1912 Einstein had made a business trip from Prague to Berlin. On that occasion he had visited Elsa. From then on a clandestine correspondence between them developed which became increasingly affectionate, if not loving, with hints of intimacy, and in which Einstein made no bones about the burdens of his current family life.[81] It appears to me that these contacts had already played a crucial role (alluded to above) in distancing Albert from Mileva in the Prague years.

At some time in 1917 Einstein became ill, suffering successively from a liver ailment, a stomach ulcer, jaundice, and general weakness. These troubles were presumably caused by intellectual overexertion combined with difficulties in procuring proper nourishment – there was a war going on. During his illness Elsa took care of her cousin. In the summer of 1917 Albert moved into an apartment next to hers. Towards the year's end he wrote to a friend, 'I have gained four pounds since last summer, thanks to Elsa's good care. She herself cooks everything for me.'[82] Nevertheless he was bedridden for several months during 1918. It was during that year that he and Elsa decided to get married – which of course demanded legal dissolution of his marriage to Mileva, still formally in force.

Divorce procedures were now set in motion, as can be seen from a handwritten draft agreement between the two parties, dated 12 June 1918, and signed by Einstein. On 20 November a court hearing took place in Zürich[83] at which neither party appeared. Mileva was represented by counsel Dr Emil Zürcher, who stated that he was a friend both of Mileva, complainant, and Albert, defendant. Counsel stated at the hearing that he was in possession of a letter by the defendant, dated 31 August 1918, in which the defendant admitted adultery. (There is no mention when that began.) Counsel's request for more detailed information had remained unanswered, obviously out of fear that the lady in question might be called to court in Berlin. Counsel could state, however, that it concerned a cousin of the defendant, a resident of Berlin, 'a person of the better classes'. Counsel had further received a postcard and a telegram from the defendant requesting that the procedures be expedited, and noted that the defendant's deposition could be considered reliable since he was 'one of the world's most famous scholars'.

On 14 February 1919 the same Zürich court ratified the decree of divorce. The stipulations specified that the husband should provide child support, and that he should have the right to have the sons with him during school holidays. The wife was to be responsible for the care and education of the children. In addition Einstein was ordered to deposit 40,000 German marks in a Swiss bank, the interest on which was to be at Mileva's disposal. He was moreover to be obliged to transfer to her the monetary award accompanying the Nobel prize, if and when he should receive that prize. Finally, he would not be allowed to marry for a period of two years – a clause that could be legally binding only in Switzerland.

I have seen the canceled Nobel prize check for 121,572.43 Swedish kronor (about $32,000 at the exchange rate of that time) which in 1923 Einstein indeed transferred to Mileva.[84] From that money she bought three houses in Zürich. Two were sold rather soon. She lived in the third, at Huttenstrasse 62, for most of the rest of her life.[85]

After the divorce Mileva had initially taken on her own family name, Marity, but by decree of the cantonal government of Zürich, dated 24 December 1924, she was given permission to revert to the name Einstein.

Late in life Einstein wrote of Mileva: 'She never reconciled herself to the separation and the divorce, and a disposition developed reminiscent of the classical example of Medea. This darkened the relations to my two boys, to whom I was attached with tenderness. This tragic aspect of my life continued undiminished until my advanced age.'[86]

Albert and Elsa were married on 2 June 1919. He was forty, she forty-three. Elsa's apartment became their home. During the following years she accompanied her husband on several journeys abroad, to the United States and to the Far East.

Einstein's mother had always remained opposed to Mileva. She was quite happy with her son's new union, however. Six months after Albert's and Elsa's marriage she came to their home to spend her last days. She was terminally ill with abdominal cancer.

Elsa's response to her second marriage has been described by Charlie Chaplin, who first met her in Pasadena in 1931: 'She was a square-framed woman with abundant vitality; she frankly enjoyed being the wife of the great man and made no effort to hide the fact; her enthusiasm was endearing.'[87]

Einstein may well have enjoyed being taken good care of and having a home in which he could receive friends. Yet it is hard to escape the impression that at deeper levels the marriage was a mismatch. A friend of his has given this picture: 'He, who had always had something of the bohemian in him, began to lead a middle-class life . . . in a household such as was typical of a well-to-do Berlin family . . . in the midst of beautiful furniture, carpets, and pictures . . . When one entered . . . one found Einstein still remained a "foreigner" in such a surrounding – a bohemian guest in a middle-class home.'[88] Nor does one have a sense of much intimacy between Elsa and her husband. The bedroom next to hers was occupied by her daughters; his was down the hall.[89] They do not appear to have been a couple much given to joint planning and deliberation. 'Albert's will is unfathomable,' Elsa once wrote.[90]

On various occasions Einstein expressed reservations about the bliss attendant on the holy state of matrimony. When during his and Elsa's visit to Japan in 1922, only three years after they had been married, someone observed him incessantly cleaning his pipe and asked him whether he smoked for the pleasure of smoking or in order to unclog and refill his pipe, Einstein replied: 'My aim lies in smoking, but as a result things tend to clog up, I'm afraid. Life, too, is like smoking, especially marriage.'[91] I know for a fact that in the early 1920s Einstein developed a strong attachment to a young woman. This extramarital affair came to an end in 1924. Several others appear to have occurred in later years.[89]

When in 1933 the Nazis made it impossible for the Einsteins to stay in Berlin they emigrated to the United States, where they settled in Princeton, New Jersey. In 1934 Ilse, Elsa's eldest daughter died in Paris after a painful illness. Elsa never recovered from this loss. Shortly after she became gravely ill and died of heart disease in 1936.

In 1939 Maja came to live with her brother in Princeton.

Since 1922, the year of her husband's retirement, the couple had lived on a small estate in Colonnata, outside Florence, bought for them by Einstein. In 1939, Mussolini's racial laws forced Maja to leave Italy. Her husband, non-Jewish, moved in with friends in Geneva. The Second World War prolonged her stay in the United States. After hostilities had ended she began making preparations for rejoining her husband. It was not to be. In 1946 she suffered a stroke and remained bedridden thereafter. Her situation deteriorated; in the end she could no longer speak, though her mind remained clear. Every night after dinner, Einstein would go to the room of his sister, who was so dear to him, and read to her.

She died in the Princeton home in June 1951.

The story of Mileva's later life is one of misfortune and despondency.[92] Her greatest sorrow was the severe chronic illness of her younger son (see the next section), who lived with her till her death. His care and frequent hospitalization created great financial burdens. Mileva frequently complained that Einstein did not help sufficiently. She made some money herself by giving piano lessons and private instruction in mathematics.

Her own health was not good either. She suffered from sclerosis and had several light strokes requiring hospitalization. During her last years she became paranoid, suspecting that people were out to steal her possessions. As Helen Dukas has told me, after her death 80,000 Swiss francs were found hidden in her mattress.

In May 1948 Mileva was taken to a private clinic. A stroke had left her paralysed on the left side. She could still talk but was confused. Her son and Lisbeth Hurwitz came for visits.

On 4 August 1948 her tragic life came to an end. She was buried in Zürich's Nordheim cemetery. A Russian priest performed the obsequies. The newspaper announcement of Mileva's death[93] was signed by Hans Albert and his wife and by Eduard – not by her former husband.

Finally I must reiterate that at no time in her life did Mileva herself lay any claim to shared fame with Einstein, as others, with so little taste and understanding, have seen fit to do in her behalf.

8. Tete

In their childhood Einstein's sons visited Mileva's family several times. Her mother always called Eduard by the pet name 'Dete', which in her language means child. His elder brother Hans Albert pronounced this as 'Tete'. That was the name by which he remained known, by his parents, by Lisbeth Hurwitz in her diaries, and by many others.[94]

Tete learned to read early, and had almost uncanny gifts of retention of what he had read. At the age of six he had already gone through Hauff's fairy tales.

On 13 August 1915 (when Tete was five), Lisbeth Hurwitz noted in her diary: 'The little one is so lively and cheerful.'[95] The relations with the mother, now separated, left much to be desired. In 1916 Mileva wrote to a friend: 'To my deepest regret I have noticed that my children do not understand me and bear a kind of grudge against me. Although it is painful, I believe it would be better for their father if he would not see them any more.'[96]

In the spring of 1917 Tete started school in Zürich. Two years later, at the age of nine, he was already reading Goethe, Schiller, and other German classics.[97]

Tete was not quite four years old when in 1914 his parents went their separate ways. In the summer of 1919 Einstein came for a family visit, the first of several he was to make after his divorce. He, Mileva, and the boys went to Arosa where Tete had to stay in a sanatorium for some time. From early childhood he had health problems, suffering in particular from severe pains in the ears and from headaches.

A year later the boys went for the first of a number of family visits in Berlin. Tete idolized his father, yet from his side a love–hate relation had begun to develop. Trbuhović ascribed this to Tete's strong need for recognition.[98] In the book by Tete's fellow-students one finds the comment: 'It must seem absurd to his classmates when in her book Mrs. Trbuhović speaks[98] of a "thirst for glory of an ambitious boy".'[99]

The high-school years, 1923–9, were the best period in Tete's life. Nothing is said about that time in Trbuhović's book. According to her, Tete increasingly showed inclinations to studies in the sciences.[100] According to his schoolmates, 'he had little interest in scientific subjects.'[101]

In high school Tete wrote prolifically, verse as well as prose. Many of these writings are reproduced in his schoolmates' book. He was particularly given to producing aphorisms, most of them written for the school paper; in 1931 some of the best were published in the *Neue Schweizer Rundschau*. It was only natural that in 1925 he would choose the literary branch of high school to continue his education. Some subjects, notably natural sciences and Greek, bored him.

Here follow recollections of three schoolmates about Tete's writing. 'He was above all interested in the description of complex human relations. He had an unheard-of grasp of German literature. He mastered the German

Einstein with his sons Hans Albert and Eduard, mid 1920s. (©Evelyn Einstein.)

language so well that he wrote brilliantly. In our school years, Eduard had a pronounced sense for humor and comic situations. His laughter, also about himself, was heart-warming . . .[102] He was a splendid schoolmate . . . He did not show the dark sides of his psyche which already then were certainly present . . .[103] We have mostly experienced Edi as vivacious, imaginative, inventive, productive, often humorous, and time and again as brilliant.'[104]

In those years Tete also continued to read copiously. He was familiar with the modern authors of the time: Kafka, Strindberg, Morgenstern, Stefan George, and especially Rilke. Once he gave a one-hour lecture in class about Rilke, on which his teacher commented that it had taught him much. He also studied the writings of Freud.[105]

From his youth Tete had shown himself to be musical. He played the piano very well. According to a schoolmate: 'Edi's enthusiasm for playing has stayed in my memory. All scepticism, insecurity, irony, and the slight sense of being absent that struck me about Edi in school vanished completely while he played. At the piano Edi appeared to me to be a quite the 6 January 1928 issue of the Swiss satirical weekly *Nebelspalter* he devoted a wonderfully caustic poem to the atonal compositions of the Viennese school, for which he did not care.

In 1926 Tete joined the voluntary military preparatory education, a group of high-school students that gathered on Saturday afternoons. In addition they occasionally went to camp. Tete showed endurance during those outings in spite of his corpulence. During a festive evening he read several of his satirical verses devoted to these exploits.

At the final high-school examination Tete was placed second, with highest possible marks in German and Latin.* All in all he had done very well in the *Gymnasium* years. He had made friendships, satisfied his teachers, and had developed a promising personal lifestyle.

Not long thereafter he became severely mentally ill.

The preceding years had shown no clear evidence of what was now to come. The only relevant comment by a school comrade[108] was that he had noticed how Tete often had cotton in his ears, obviously because of the severe earaches mentioned earlier.

After finishing high school Tete paid another visit to his father in Berlin. Thereafter he enrolled in the University in Zürich to study medicine, hoping to specialize later in psychiatry. He managed to go through three semesters, participating also in student feasts, to which he once again contributed verses written for the occasion.

The genesis and precise nature of Tete's illness are unclear to me. Helen Dukas once told me that Einstein recognized rather early signs of dementia praecox, but I do not know how early. A biographer has written: 'Eduard inherited from his father the facial traits and the musical talents, from his mother the tendency to melancholy.'[109] It is painful to recognize in Tete's own aphorisms of 1931 forebodings of his fate: 'Man only loves destiny. It is the worst destiny to have no destiny and to be no destiny for anyone else . . .'[110]

Manifestations of disturbances became more and more evident soon after he had left high school. He covered the walls of his room with pornographic pictures, his passion changed from books to women.[111] He had attacks of rage so violent that he had to be taken to Burghölzli, a psychiatric institution near Zürich. When afterwards he attended university he was discretely accompanied by a male nurse.[112]

After his early efforts to attend university Tete had to give up. Much of the time he was in a state of lethargy. On 1 July 1934, Lisbeth Hurwitz wrote in her diary that Tete had become fat, was restless, and read biographies of great men and obscene plays.[113] On 13 June 1937 she wrote that he was in bad shape, rarely left the house, and had screaming attacks to deaden the earaches. His periods of institutionalization in Burghölzli became increasingly longer.

* The complete list of his marks is found in ref. 107.

Fellow-students have described later encounters on the street with him. One remembered trying unsuccessfully to talk to him. 'It was clear that he lived in another world and felt bothered by me.' Another: 'He was talking to the sky and quickly walked on when he saw me.' Old friends who visited him in Burghölzli felt that Tete was not happy with such encounters.[114]

A person who saw him two years before his death in Burghölzli has written[115] that he met Tete outdoors where he had been working in the fields. He was rather fat, very pale, and looked almost frighteningly like his father. His conversation was schizophrenic. He did not like the outdoor work but realized that it was good for him.[115]

After Mileva's death Tete continued to live in Burghölzli for another seventeen years, during which he never mentioned his mother. He died on 25 October 1965, and was buried in the Hönggerberg cemetery.

The newspaper announcement[116] of Tete's death, signed by his brother Hans Albert and stepsister Margot, mentions that Eduard Einstein was the 'son of the deceased Professor Albert Einstein.' I find it scandalous that no mention was made of Mileva.

9. Conclusion

In 1936, shortly after Elsa's death in Princeton, Einstein wrote to his colleague and friend Max Born: 'I have acclimated extremely well here, live like a bear in a cave, and feel more at home than I ever did in my eventful life. This bearlike quality has increased because of the death of my comrade [*Kameradin*], who was more attached to people [than I].'[117]

In 1952 he wrote to his friend Jakob Ehrat, a fellow-student from the ETH days: 'I am so far all right in that I have victoriously survived the Nazi-time and two wives.'[118]

One month before Einstein's death, Michele Besso, a friend since the days at the Patent Office, passed away, whereupon Einstein wrote to his widow: 'What I most admired in him as a human being is the fact that he managed to live for many years not only in peace but also in lasting harmony with a woman – an undertaking in which I twice failed rather disgracefully.'[119]

Einstein wrote about his marriages with frankness and insight, though not always with grace. He had not been a good husband, and he knew that. He had been a better father, though I would not venture to say how good – especially because it has so far remained unclear what became of Lieserl.

The reason for lack of strength in Einstein's family ties is, I think, obvious. To be creative in establishing lasting deep human relations demands efforts that Einstein was simply never willing to make. His full creative exertions went completely and always into science, perhaps at a cost to himself, certainly at a cost to those close – or trying to be close – to him.

Einstein's keen perception (with which I am familiar from numerous talks with him) of human qualities and frailties extended to an awareness of his own personal demons. As he once said: '*Ich muss in den Sternen suchen was mir auf Erden versagt ist:*' – 'I must search in the stars for what is denied me on earth . . .'

References

1. A. Pais, *Subtle is the Lord*, Clarendon Press, 1982; referred to below as *SL*.

2. D. Trbuhović, *Im Schatten Albert Einsteins*, 4th edn, Verlag Paul Haupt, Bern and Stuttgart, 1988.

3. *Collected Papers of Albert Einstein*, Vol. 1, ed. J. Stachel *et al.*, Princeton University Press, 1987; referred to below as *CP*. This volume contains numerous letters exchanged between Einstein and Marić. These have meanwhile appeared in English translation; J. Renn and R. Schulmann, *Albert Einstein–Mileva Marić, the love letters*, Princeton University Press, 1992 (referred to below as RS).

4. E. Rübel, *Eduard Einstein*, Verlag Paul Haupt, Bern and Stuttgart, 1986.

5. Pauline Einstein, letter to her mother, 1 August 1886, *CP* Vol. 1, doc. 2, p. 3.

6. Reprinted in *CP* Vol. 1, p. xlviii.

7. A. Einstein, *Helle Zeit, dunkle Zeit*, p. 9, ed. C. Seelig, Europa Verlag, Zürich, 1956.

8. For the complete list of grades see *CP* Vol. 1, p. 16.

9. See *SL*, Chapter 3, for more details about Einstein's earlier years.

10. The information on Miloš's later years was gathered by Professor Zimmermann; see ref. 2, pp. 161, 162.

11. This account of Mileva's education is taken from ref. 2.

12. Ref. 2, pp. 60, 61.

13. *CP* Vol. 1, doc. 67, 27 July 1900.

14. After 20 October 1897, *CP* Vol. 1, doc. 36, p. 58, RS doc. 1.

15. 16 February 1898, *CP* Vol. 1, doc. 39, p. 211. Only an envelope remains of an earlier letter, see same page; also RS doc. 2.

16. 16 April–8 November 1898, *CP* Vol. 1, doc. 40, p. 213, RS docs. 3–5.

17. 13 or 20 March 1899, *CP* Vol. 1, doc. 45, p. 215, RS doc. 6.

18. Early August 1899, *CP* Vol. 1, doc. 50, p. 220, RS doc. 7.

19. 10? August 1899, *CP* Vol. 1, doc. 52, p. 225, RS doc. 8.

20. Between 10 August and 10 September 1899, *CP* Vol. 1, doc. 53, p. 228, RS doc. 9.

21. 10 September 1899, *CP* Vol. 1, doc. 54, p. 229; 28? September 1899, *ibid.*, doc. 57, p. 233, RS docs. 10, 11.

22. 10 October 1899, *CP* Vol. 1, doc. 58, p. 237, RS doc. 12.
23. Probably 1900, *CP* Vol. 1, doc. 61, p. 242, RS doc. 13.
24. See, e.g., *CP* Vol. 1, doc. 49.
25. 29? July 1900, *CP* Vol. 1, doc. 68, p. 248, RS doc. 14.
26. 6 August 1900, *CP* Vol. 1, doc. 70, p. 251, RS doc. 16.
27. 9? August 1900, *CP* Vol. 1, doc. 71, p. 252, RS doc. 17.
28. For more on this period see *SL*, pp. 45, 46.
29. 19 September 1900, *CP* Vol. 1, doc. 76, p. 261, RS doc. 22.
30. 3 October 1900, *CP* Vol. 1, doc. 79, p. 266, RS doc. 23.
31. Second half of May? 1901, *CP* Vol. 1, doc. 107, p. 300, RS doc. 33.
32. 15 April 1901, *CP* Vol. 1, doc. 101, p. 291; also 12 December 1901, *ibid.*, doc. 127, p. 322, RS docs. 28, 45.
33. 27 March 1901, *CP* Vol. 1, doc. 94, p. 281, RS doc. 25.
34. About 1 September 1900, *CP* Vol. 1, doc. 74, p. 257, RS doc. 20.
35. 13? September 1900, *CP* Vol. 1, doc. 75, p. 259, RS doc. 21.
36. 31? July 1901, *CP* Vol. 1, doc. 121, p. 313, RS doc. 41.
37. Letter to Helene Savić, November–December 1901, *CP* Vol. 1, doc. 125, p. 319.
38. 28 November 1901, *CP* Vol. 1, doc. 126, p. 320, RS doc. 44.
39. 14? August 1900, *CP* Vol. 1, doc. 72, p. 254, RS doc. 18.
40. About 1 September *CP* Vol. 1, doc. 74, p. 257, RS doc. 20.
41. 10 April 1901, *CP* Vol. 1, doc. 97, p. 286, RS doc. 27.
42. 30 April 1901, *CP* Vol. 1, doc. 102, p. 293, RS doc. 29.
43. May 1901, *CP* Vol. 1, doc. 107, p. 300, RS doc. 33.
44. May 1901, *CP* Vol. 1, doc. 108, 301, RS doc. 44.
45. 8 July 1901, *CP* Vol. 1, doc. 116, p. 310, RS doc. 39.
46. 28 May 1901, *CP* Vol. 1, doc. 111, p. 304, RS doc. 36.
47. 4? June 1901, *CP* Vol. 1, doc. 112, p. 306, RS doc. 37.
48. 7? July 1901, *CP* Vol. 1, doc. 114, p. 308, RS doc. 38.
49. 22 July 1901, *CP* Vol. 1, doc. 119, p. 312, RS doc. 40.
50. 12 December 1901, *CP* Vol. 1, doc. 127, p. 322, RS doc. 45.
51. 17 December 1901, *CP* Vol. 1, doc. 128, p. 325, RS doc. 46.
52. 19 December 1901, *CP* Vol. 1, doc. 130, p. 328, RS doc. 47.
53. 28 December 1901, *CP* Vol. 1, doc. 131, p. 329, RS doc. 48.
54. 4 February 1902, *CP* Vol. 1, doc. 134, p. 332, RS doc. 49.
55. Pauline Einstein, letter to Pauline Winteler, 20 February 1902, *CP* Vol. 1, doc. 138, p. 336.
56. Helen Dukas, private communication.
57. Reproduced in ref. 2, p. 80.
58. M. Flückiger, *Albert Einstein in Bern*, p. 133, Paul Haupt, Bern, 1974.

59. A. Einstein, letter to C. Seelig, 5 May 1952, reproduced in C. Seelig, *Albert Einstein*, Europa Verlag, Zürich, 1960.

60. M. Einstein, *Beiträge zur Überlieferung des Chevalier du Cygne und der Enfance Godefroi*, Druck, Erlangen, 1910.

60*a*. For her reminiscences see E. Roboz Einstein, *Hans Albert Einstein*, Iowa Inst. of Hydraulic Research, University of Iowa, 1991.

61. *New York Times*, 27 July 1973.

62. H. A. Einstein, letter to A. Einstein, written about 1933.

63. H. A. Einstein, letter to C. Seelig, 18 April 1955; ETH Bibliothek Zürich, HS 304:566.

64. 19? September 1903, *CP* Vol. 5, doc. 13, RS doc. 54.

65. Ref. 2, p. 58. 66. Ref. 2, p. 87. 67. Ref. 2, p. 90. 68. Ref. 2, p. 97.

69. R. Schulmann, letter to A. Pais, 11 July 1990.

70. A. Joffe, Pamiati Alberta Einsteina, *Uspekhi Fys. Nauk*, 57, 187, 1955.

71. Ref. 2, p. 95. 72. Ref. 2, p. 108.73. Ref. 2, p. 111.74. Ref. 2, pp. 108, 115.

75. Ref. 59, p. 203.76. Ref. 2, p. 111. 77. Ref. 2, pp. 120, 121.

78, Mileva to Helen Savić, 12 March 1913, quoted in ref. 2, p. 122.

79. Ref. 2, p. 126.

80. For details of events leading to this move see *SL*, Chapter 14, section (a).

81. See *CP*, Vol. 5, currently in preparation.

82. A. Einstein, letter to H. Zangger, 6 December 1917.

83. Protocol numbered 1386/1918 of the *Bezirksgericht Zürich*, II. *Abteilung*, which includes the draft agreement of 12 June; copy in the Staatsarchiv des Kantons Zürich.

84. H. Dukas, private communication.

85. Ref. 2, pp. 180, 182.

86. A. Einstein, letter to C. Seelig, 5 May 1952.

87. Charles Chaplin, *My autobiography*, p. 346, The Bodley Head, London, 1964.

88. P. Frank, *Einstein, his life and times*, p. 124, Knopf, New York, 1953.

89. F. Herneck, *Einstein privat*, Der Morgen, Berlin, 1978.

90. E. Einstein, letter to P. Ehrenfest, 5 April 1932.

91. J. Ishiwara, *Einstein ko en roku*, p. 193, Tokyo-Tosho, Tokyo, 1978.

92. See esp. ref. 2, pp. 195–202.

93. Repr. in ref. 2, p. 198.94. Ref. 2, p. 125. 95. Ref. 2, p. 132.

96. Letter to Helene Savić, 8 September 1916, reproduced in ref. 2, p. 139.

97. Ref. 4, p. 11. 98. Ref. 2, pp. 147, 148.99. Ref. 4, p. 23.

100. Ref. 2, p. 147.

101. Ref. 4, p. 23. 102. Ref. 4, p. 36. 103. Ref. 4, p. 53. 104. Ref. 4, p. 61.

105. Ref. 4, pp. 57–60. 106. Ref. 4, p. 95.107. Ref. 4, p. 78.108. Ref. 4, p. 57.

109. C. Seelig, *Albert Einstein*, (ref. 59), p. 192.

110. Ref. 4, p. 81.

111. Ref. 2, p. 169, ref. 4, p. 111.

112. Ref. 4, p. 102.

113. Ref. 2, p. 189.

114. Ref. 4, p. 113.

115. 'M. W.' in *Brückenbauer*, 19 November 1965.

116. Repr. in Ref. 2, p. 204.

117. A. Einstein, letter to M. Born, undated, probably 1937; reproduced in *Einstein–Born Briefwechsel*, p. 177, Nymphenburger, Munich, 1969.

118. A. Einstein, letter to J. Ehrat, 12 May 1952.

119. A. Einstein, letter to V. Besso, 21 March 1955, reproduced in *Einstein–Besso Correspondence 1903–1955*, p. 537, ed. P. Speziali, Hermann, Paris, 1972.

2

Reflections on Bohr and Einstein

To set the stage I shall first relate three stories widely separated in time.

In 1923 Max Born, a distinguished physicist from the Bohr–Einstein generation, submitted a letter to the Göttingen Academy of Sciences in which he proposed both Einstein and Bohr for nomination as foreign members. In recommending Bohr he wrote, 'His influence on theoretical and experimental research of our time is greater than that of any other physicist.'[1] Note that Born's personal relations with Einstein were closer than with Bohr. At about that same time Percy Bridgman from Harvard wrote to an acquaintance that Bohr was now idolized as a scientific god throughout most of Europe.[2]

In 1963 Heisenberg, a member of the next generation, wrote in an obituary to Bohr, 'Bohr's influence on the physics and the physicists of our century was stronger than that of anyone else, even than that of Albert Einstein.'[3]

My third story concerns a discussion about Bohr which took place in the early 1980s between myself and a friend of mine, one of the best and best-known physicists of my own generation, the one following Heisenberg's.

'You knew Bohr well,' he said.

'I did,' I replied.

'Then tell me,' he asked, 'what did Bohr really do?'

'Well,' I replied, 'first and foremost he was one of the founding fathers of the quantum theory.'

'I know,' he answered, 'but that work was superseded by quantum mechanics.'

'Of course,' I said, and then proceeded to explain Bohr's role in quantum mechanics, in particular his introduction of complementarity. This, I found out, had not been clear to my friend.

What did Bohr really do? Why is it that complementarity, which he himself considered his main contribution, is not mentioned in some of the finest physics textbooks, such as the one on quantum mechanics by Dirac, the historically oriented quantum mechanics text by Tomonaga, or the lectures by Feynman? What, for that matter, did Einstein really do?

It has taken me the writing of two biographies, one on Einstein and one on Bohr, each nearly 600 pages long, to give to the best of my ability an answer to all these questions. I shall try to summarize my views in this brief chapter, which focuses on the personal and intellectual contact between these two men.

Einstein and Bohr are two of the three physicists without whom the birth of that uniquely twentieth-century mode of thought, quantum physics, is unthinkable.

The three are, in order of appearance: Planck, the reluctant revolutionary, discoverer of the quantum theory who did not at once understand that his quantum law meant the end of an era now called classical.

Einstein, discoverer of the quantum of light, the photon, who at once realized that classical physics had reached its limits, a situation with which he never could make peace.

And Bohr, founder of the quantum theory of the structure of matter, also immediately aware that his theory violated sacred classical concepts, but who at once embarked on the search for links between the old and the new, achieved with a considerable measure of success in his correspondence principle.

How different their personalities were.

Planck, in many ways the conventional university professor, teaching his courses, delivering his Ph.D.s.

Einstein, rarely lonely, mostly alone, who did not really care for teaching classes, and who never delivered a Ph.D. He did have collaborators: I have in fact counted more than thirty physicists with whom he published jointly. Nevertheless it was his deepest need to think separately, to be by himself.

And Bohr, always in need of other physicists especially young ones, to help him clarify his own thoughts, always generous in helping them clarify theirs, not a teacher of courses nor a supervisor of Ph.D.s but always giving inspiration and guidance to postdoctoral and senior research.

It is essential for what follows to recall that the evolution of the quantum theory is divided into two sharply distinct periods.

The first, from 1900 to 1925, now known as the time of the old quantum theory, spanned the most unusual years in all the history of physics. Quantum laws and regularities were discovered which, experiment showed, were to be taken most seriously, yet which violated the fundamental logic on which the physics of that period rested. A prime example of this bizarre state of affairs is Bohr's work, beginning in 1913, which for the first time made atomic structure into a subject of scientific inquiry. How new that development was can be appreciated by recalling the situation at the turn of the century, as described by Andrade: 'It is perhaps not unfair to say that for the average physicist at the time, speculations about atomic structure were something like speculations about life on Mars – very interesting for those who like this kind of thing but without much hope of support from convincing scientific evidence and without much bearing on scientific thought and development.'[4]

The best characterization of Bohr's activities during those years was given in 1949 by the seventy-year-old Einstein: 'That this insecure and contradictory foundation was sufficient to enable a man of Bohr's unique instinct and tact to discover the major laws of the spectral lines appeared to me as a miracle – and appears to me as a miracle even today. This is the highest sphere of musicality in the sphere of thought.'[5] Those years of struggle in the *clair-obscur* left an indelible mark on Bohr's style, once again best expressed by Einstein: 'He utters his opinions like one perpetually groping and never like one who believes to be in the possession of definite truth.'[6] As Bohr himself often used to say, 'Never express yourself more clearly than you think.'

What was Einstein up to in that first quarter of the twentieth century? Of his many extraordinary contributions I single out his relativity theories, mainly in order to stress a counterpoint with quantum physics.

Einstein's first paper on special relativity, published in 1905, is axiomatic in structure; the whole edifice is erected on new first principles. It is so perfectly written that what remained to be done ever after was to work out further consequences of the Einstein postulates. Not one word in the paper needs to be changed in the light of later developments.

In 1916 Einstein published the first correct version of the general theory of relativity, the new theory of gravitation, arguably the most profound contribution to the science of our century. That paper too has a high level of perfection but not all its fundamental consequences were at once understood. The main reason is that at first Einstein and others considered only the implications for relatively weak gravitational fields. The consequences for strong gravitational fields, gravitational collapse, black holes, and the history of the early universe were recognized as subjects of major importance only after the deaths of both Einstein and Bohr. In the intervening years fine work on relativity was done and significant progress was made but the number of physicists so engaged was quite limited.

By contrast Bohr's papers on the structure of atoms unleashed a veritable flood of activity in many research centers, including contributions by Bohr himself. I wish to keep this account non-technical but am nevertheless compelled to recite the successes of the old quantum theory, simply because these are not all that well known even to professional physicists. My list of achievements includes the discoveries of the principal, angular momentum, and magnetic quantum numbers, as well as their selection rules; the theory of the linear Stark effect; Bohr's work on the ground states of complex atoms, which laid the foundations of quantum chemistry, which influenced Pauli to introduce the exclusion principle, and which, in turn, led Uhlenbeck and Goudsmit to the discovery of electron spin. All this, as well as the discoveries of Bose–Einstein and Fermi statistics belongs to the era of the old quantum theory. In retrospect these many successes are all the more fabulous and astounding because they are based on analogies – atomic orbits that are similar to the motions of planets around the sun; spin that is similar to planets that rotate while orbiting – which are in fact false. No wonder then that the old quantum theory also showed capital flaws, most notably its failure to explain all atomic spectra beyond the simplest one, that of hydrogen.

At least as important, perhaps even more so, as Bohr's participation in this research was his emergence as the leader of the field. This was the result, not only of his stature in science, but also of his gifts and his need to guide the work of others. In 1921 Sommerfeld called him 'the director of atomic physics'. That was the year in which his Copenhagen institute began operations. From the start and during the next two decades it was to be the world's leading center in theoretical physics.

By 1921 Einstein's fame had already reached mythical proportions. Yet neither his relativity theories nor his personality lent themselves to stimulate such massive activities as were unfolding in quantum physics. All of which goes to explain what Born wrote in 1923; also in part what Heisenberg wrote in his obituary; and also in part what my friend was unaware of in the 1980s.

So much for the period up to 1925. In that year the new physics, the new logic, emerged with the discovery of quantum mechanics. I shall turn presently to what happened thereafter.

I have been privileged to know both Bohr and Einstein personally. This came about as follows. When after the Second World War I emerged from hiding in my native Holland I wanted to continue my studies abroad. So I applied for fellowships to Bohr in Copenhagen and to Pauli, who was at the Institute for Advanced Study in Princeton at that time. I received a Rask Ørsted Fellowship from Denmark and was also accepted in Princeton. I decided to go to Copenhagen first.

So it came about that I first met Bohr when I went to Copenhagen in January 1946 as the first of the post-war postdoctoral crop from abroad. Some months later Bohr asked me whether I would be interested in working together with him day by day for the coming months. I was thrilled and accepted. The next morning I went to his home at Carlsberg. The first thing Bohr said to me was that it would be profitable to work with him only if I understood that he was a dilettante. The only way I knew to react to this unexpected statement was with a polite smile of disbelief. But evidently Bohr was serious. He explained how he had to approach every new question from a starting-point of total ignorance. It is perhaps better to say that Bohr's main strength lay in his formidable intuition and insight, rather than in erudition. Before long I caught the spirit of his disinterested, joyously obsessive pursuit of truth, especially during the summer of 1946, which I spent with Bohr and his family in their summer home in Tisvilde. There I got to know well his wife Margrethe, a charming and formidable lady, and the sons, Aage, Hans, Erik, and Ernest. We have been friends ever since.

In September 1946 I went to Princeton. The first thing I learned was that, in the meantime, Pauli had gone to Zürich. Bohr also came to Princeton that month for the Princeton University bicentennial meetings, which we both attended. There came a day when Bohr said to me: '*Nu skal vi hilse paa Einstein*,' 'Now let us say hello to Einstein.' Thus came about my first meeting with Einstein, who greeted a rather awed young man with a friendly smile and outstretched hand. The conversation turned to the quantum theory. I listened as the two of them argued. I recall no details but remember distinctly my first impressions. They liked and respected each other. With a fair amount of passion they were talking past each other. From many previous discussions I could follow Bohr's reasoning, but I did not understand what Einstein was talking about.

Not long thereafter I encountered Einstein in front of the Institute, told him that I had not followed his argument with Bohr, and asked if I could come to his office some time for further enlightenment. He invited me to walk home with him.

Thus began a series of discussions, always in German, that continued for nine years, until shortly before his death. I would visit him in his office or accompany him on his lunchtime walk home. Less often I would visit him there. In all I saw him about once every few weeks.

Whenever I met Einstein our conversations might range far and wide, touching on politics, the bomb, the Jewish destiny, or on less weighty matters. Invariably, however, the discussion would turn to physics, particularly to the interpretation of quantum mechanics, which Einstein did not cease pondering as long as he lived. He was explicit in his opinion that the most commonly held view on this subject could not be the last word. It did not take long before I had a grasp of the points that Einstein had debated with Bohr on that first occasion I had seen them together. I had of course already heard from Bohr something about Einstein's views. Now I became directly exposed to the details of the latter's arguments.

In order to give a first glimpse of Einstein's thinking I shall relate what happened during one of our lunchtime walks. It must have been around 1950. At one point Einstein suddenly stopped, turned to me, and asked me if I really believed that the moon exists only if I look at it. The nature of our conversation was not particularly metaphysical. Rather we were discussing what is doable and knowable in the sense of physical observation, the central epistemological issue of quantum mechanics. We walked on and continued talking about the moon and the meaning of the expression 'to exist' as it refers to inanimate objects. When we reached 112 Mercer Street, I wished him a pleasant lunch, then returned to the Institute. As had been the case on many earlier occasions, I had enjoyed the walk and felt better because of the discussion even though it had ended inconclusively. I was used to that by then, and as I walked back I wondered once again, why does this man, who contributed so incomparably much to the creation of modern physics, remain so attached to the nineteenth-century view of causality? What prevents him from accepting Bohr's complementarity?

This is a natural point for me to comment on Bohr's and Einstein's personalities. I do so next. I shall then say something about their various encounters, and conclude by sketching the substance of the Bohr–Einstein dispute on the foundations of quantum mechanics.

Bohr and Einstein were in their early and late sixties respectively when I first met them. Since I am one (perhaps the last) of those who knew both men rather well personally in their later years it is hardly surprising that I have been asked off and on how, in my view, they compared. That question used to make me mildly uncomfortable, and in the past I have tended to respond evasively, simply because the better one believes one has understood people the more superficial and hopeless one's comparisons tend to become. That remains true today. I must admit, however, that a comparison between Bohr and Einstein is far more interesting than similar more frivolous debates often indulged in by physicists, a competitive breed, as to who is smarter, A or B. After all we are dealing here with men who were arguably the two leading figures in science in this century. So, abandoning my reservations, I shall reflect on what kinds of men they were.

First, physics, by which both Bohr and Einstein were possessed if not obsessed. Both would speak with intense enthusiasm and optimism about work they were engaged in. Both had enormous powers of concentration.

Both realized quite early not only the importance but also the paradoxes resulting from Planck's discovery of his radiation law. In younger years Einstein's spectrum of scientific activities was broader than Bohr's. Also in their younger years both had an urge for doing experiments, at which Bohr did better. Bohr published 200 papers in scientific journals, Einstein 270 (both numbers are approximate). All their respective most important papers appeared under their own name only. Both men were indefatigable workers, driving themselves on occasion to states of exhaustion which would lead to illness, more serious in the case of Einstein. Both taught courses in their younger but not in their later years. As mentioned above, neither had his own Ph.D. students. Neither experienced difficulty or pain in admitting to himself, if not to others, that occasionally he had been on a wrong scientific track. Neither was in the least overwhelmed by the medals, prizes, honorary degrees, and other distinctions showered upon them; their prime concern was always with what they did not understand rather than with past achievements.

Their life-spans were almost identical. Bohr lived to be 77, Einstein 76. Both chose to be cremated. Their fathers, who died relatively young, also died at about the same age; Bohr's at 56, Einstein's at 55. Einstein remained scientifically active until, literally, the day he died. From the point of view of pure science Bohr was more a spectator than an actor in his later years.

Both Bohr and Einstein were areligious. Bohr left the Lutheran church in 1911, just before he got married. Einstein said that he did not believe in a God who concerned himself with the fates and actions of human beings. He would often invoke God in his spoken and written words. ('God does not play dice.') When, in the 1930s, the celebrated actress Elizabeth Bergner asked Einstein whether he believed in God, he replied: 'One may not ask that of someone who with growing amazement attempts to explore and understand the authoritative order in the universe.' When she asked, 'Why not,' he answered: 'Because he would probably break down when faced with such a question.'[7] Imagery like that would never occur to Bohr's mind.

Neither Bohr nor Einstein had overt emotional problems. Einstein's handwriting was clear, Bohr's was poor. Music was a profound necessity in Einstein's life, but not in Bohr's. Both men had remarkably gentle voices. Both felt strongly attracted to the visual arts. Both were well read outside the sciences. Both mastered foreign languages passably but not really well. Each spoke English with a distinct and lovable accent. Each had a great wit and sense of humor and occasionally liked to tell jokes.

Bohr greatly enjoyed sports, soccer in his youth, tennis later, skiing through most of his life. Einstein did not care for any such diversions. Both loved to sail. Einstein never owned or drove a car, Bohr did. As I know from experience, his driving could on occasion be a bit scary. Both traveled extensively (Einstein only when in his forties). Both were pipe smokers (in his earlier years Bohr also smoked cigarettes), though Einstein was forbidden to smoke in later years.

In regard to family, both men grew up in closely knit warm parental homes. In Bohr's case the father dominated, in Einstein's the mother. Bohr's father was an eminent scientist, Einstein's a small businessman who had to cope with a series of failures. Bohr came from a distinctly upper-class milieu; Einstein's was middle class. Bohr's two years younger brother was (after his wife) the person most close to him. There may never have been anyone to whom Einstein felt closer than his sister (his only sibling, two years younger than him). Niels Bohr's marriage to Margrethe was to both a source of great harmony, strength, and singlemindedness. Einstein was married twice (he survived both spouses), undertakings at which, in his words, he 'twice failed rather disgracefully.' He had several extramarital affairs.

The Bohrs had six sons. Einstein and his first wife had one daughter (whose later fate is unknown) before and two sons during their marriage. Einstein also had two stepdaughters as a result of his second marriage. Bohr was a family man, a wonderful, devoted father. About Einstein I am less clear in that respect; later letters indicate that relations sometimes left something to be desired. The children brought joy but also tragedy to their parents. Four of the Bohr sons engaged in distinguished careers; the same was true of Einstein's older son. But the Bohrs lost their eldest son in a sailing accident, and their youngest, Harald, contracted meningitis when he was about a year old and died young. Einstein's younger son was schizophrenic and died (at 55) in a mental hospital.

The Bohrs had eight grandsons and nine granddaughters; Einstein had two grandsons and, by adoption, one granddaughter. It belongs to my happy memories to have seen Niels sitting on the floor playing with his grandchildren.

In other personal relations or encounters neither man was swayed by class or rank. Both were quite accessible to men and women from all walks of life, when not busy otherwise. Both were shrewd observers of human nature, always friendly and courteous, but able to be sharply critical of people in more private discussions. On general social issues both men spoke up and took action on behalf of the downtrodden. Both personally met numerous leading statesmen of their time, including Winston Churchill and Franklin D. Roosevelt.

Beginning in 1914, but most especially after the Second World War, Einstein signed or was cosignatory of numerous politically oriented declarations. Bohr did so only once, in his open letter to the United Nations in 1950. Both were highly sympathetic to the cause of Israel, though not uncritically so.

Of particular interest for what follows are the two men's attitudes towards philosophy.

Einstein had a lifelong interest in philosophy. For Bohr, less well read in philosophy than Einstein, philosophizing was part of his nature from boyhood on. His first preoccupation with philosophical problems did not arise from his physical investigations but from general epistemological considerations about the function of language as a means of communicating experience. How to avoid ambiguity: that was the problem that worried Bohr. Shortly before his death he spoke about his youthful philosophical considerations. When asked how significant those were to him at that time, he replied: 'It was, in a way, my life.'[8]

Like Einstein, Bohr had little patience with and no use for professional philosophers in later life. In his own words: 'There are all kinds of people but I think it would be reasonable to say that no man who is called a philosopher really understands what one means by the complementarity description . . . The relationship between scientists and philosophers is of a very curious kind . . . The difficulty is that it is hopeless to have any kind of understanding between scientists and philosophers directly.'[8] One of Bohr's favourite anecdotes goes like this. What is the difference between an expert and a philosopher? An expert is someone who starts out knowing something about some things, goes on to know more and more about less and less, and ends up knowing everything about nothing. Whereas a philosopher is someone who starts out knowing something about some things, goes on to know less and less about more and more, and ends up knowing nothing about everything.

Neither Bohr's nor Einstein's *oeuvre* shows any influence of the writings of philosophers.

In the preceding paragraphs similarity by and large outweighed disparity. My last point of similarity could well be more illuminating than anything said before. Both had a deep need for simplicity, in thought and in behavior. Each had a lifelong boyish – not juvenile, boyish – curiosity, and pleasure in play. They took science very seriously, but to them it was ultimately a game.

In some other respects, however, Bohr and Einstein were extreme opposites. I have already mentioned their different needs in regard to contact with other physicists and have intimated their opposing views on the interpretation of quantum mechanics. To this I should now add another most significant difference. To Bohr one and only one place was home: Denmark. Einstein never fully identified with any one country or nation;

he would call himself a gypsy, or a bird of passage. He lived in – rather than visited – many places: Germany (Ulm, Munich, Berlin), Switzerland (Aarau, Bern, Zürich), Milan, Prague, and Princeton.

I turn now to Bohr and Einstein's encounters.

They first met in April 1920, when Bohr came to Berlin to give a lecture. Einstein was enchanted. Shortly afterward he thanked Bohr by letter for 'the magnificent gift from Neutralia [Denmark] where milk and honey still flow,' and continued: 'Not often in life has a person, by his mere presence, given me such joy as you did. I am now studying your great papers and in doing so – especially when I get stuck somewhere – I have the pleasure of seeing your youthful face before me, smiling and explaining. I have learned much from you, especially also about your attitude regarding scientific matters.'[9] Bohr replied: 'To me it was one of the greatest experiences ever to meet you and talk with you. I cannot express how grateful I am for all the friendliness with which you met me on my visit to Berlin. You cannot know how great a stimulus it was for me to have the long hoped for opportunity to hear of your views on the questions that have occupied me. I shall never forget our talks.'[10]

Bohr and Einstein met again in August 1920, when Einstein stopped in Copenhagen on his way back from a trip to Norway. Einstein wrote to Lorentz: 'The trip to Kristiania [Oslo] was really beautiful, the most beautiful were the hours I spent with Bohr in Copenhagen. He is a highly gifted and excellent man. It is a good omen for physics that prominent physicists are mostly also splendid people.'[11]

The next contact occurred when Bohr wrote to Einstein on the day each had been informed of his Nobel prize: 'The external recognition cannot mean anything to you . . . For me it was the greatest honor and joy . . . that I should be considered for the award at the same time as you. I know how little I have deserved it, but I should like to say that I consider it a good fortune that your fundamental contributions in the special area in which I work [i.e. the quantum theory of radiation] as well as contributions by Rutherford and Planck should be recognized before I was considered for such an honor.'[12] Einstein was in Japan at that time. On his way back he answered Bohr's letter from aboard ship somewhere near Singapore: 'I can say without exaggeration that [your letter] pleased me as much as the Nobel prize. I find especially charming your fear that you might have received the award before me – that is typically Bohr-like [*bohrisch*]. Your new investigations on the atom have accompanied me on the trip, and they have made my fondness for your mind even greater.'[13]

Einstein went to Copenhagen again in July 1923. From Bohr's later reminiscences of that visit: 'Einstein was not more practical than I and, when he came to Copenhagen, I naturally fetched him from the railway

Einstein on the Oslofjord, June 13, 1920. To his left: Professor Heinrich Goldschmidt. To his right: Mr Jakob Schetelig. (Gift to the author from Professor Otto Walaas, Oslo.)

station. We took the street car from the station and talked so animatedly about things that we went much too far past our destination. So we got off and went back. Thereafter we again went too far, I can't remember how many stops, but we rode back and forth in the street car because Einstein was really interested at that time; we don't know whether his interest was more or was less skeptical – but in any case we went back and forth many times in the street car and what people thought of us, that is something else.'[14]

In later years Bohr and Einstein did not meet often nor can their correspondence be called extensive. Yet Einstein was to play a singularly important role in Bohr's life, as we shall see shortly. They were destined to become intellectual antagonists, but that in no way diminished their mutual respect and affection. When Einstein was gone and Bohr had only one more year to live he once said: 'Einstein was so incredibly sweet. I want also to say that now, several years after Einstein's death, I can still see Einstein's smile before me, a very special smile, both knowing, humane and friendly.'[14]

I come to my last topic, the time after 1925, the year in which quantum mechanics was born. Neither Bohr nor Einstein contributed to the initial discoveries which span the period 1925–6. The impact of this development on our two heroes was profound, however. I shall first describe what it did to Einstein.

For that purpose I must first track back to the days of Newton, who had conjectured that light consists of small bullets. Almost simultaneously, in 1690, Christiaan Huyghens published his *Traité de la Lumière* in which he proposed that light consists of waves. The two theories were in clear conflict. Particles are not waves, waves are not particles. Each theory had its adherents until early in the nineteenth century Young and Fresnel showed experimentally that light exhibits interference phenomena that can be understood only on the wave theory. The verdict was clear, it seemed: Huyghens' wave picture was in, Newton's particle picture was out.

There matters stood until 1905 when young Einstein came forth with his light-quantum hypothesis: under certain circumstances monochromatic light with frequency ν behaves as if it consists of mutually independent particles, photons, with energy $h\nu$ (h is Planck's constant).

Quite understandably Einstein's photon initially encountered intense resistance. For example, as late as 1924 Bohr proposed a long-since-forgotten alternative to photons. The situation was indeed incomparably more grave than the Newton–Huyghens controversy, where one set of concepts simply had to yield to another. Here, on the other hand, it became clear, as time went by, that not only could the wave picture lay claim to success for some phenomena that excluded the particle picture, but also that the particle picture could make similar claims for other phenomena that excluded the wave picture. What was going on?

Einstein himself needed no convincing that his photons had no place in the wave theory. In 1909 he had already written: 'It is my opinion that the next phase in the development of theoretical physics will bring us a theory of light that can be interpreted as a kind of fusion of the wave and the particle theory . . . The wave structure and the quantum structure . . . are not to be considered as mutually incompatible.'[15]

In 1916 Einstein made another fundamental contribution to the quantum theory. He realized that, when an excited atom emits a photon the theory can predict neither the time at which nor the direction in which the photon is emitted. That is to say the theory violated the classical principle of causality, according to which if at a given time an isolated system is in a fully specified state then one should be able to predict rigorously its behavior at any later time. That state of affairs troubled Einstein greatly, then and later.

As a final comment on Einstein's role prior to quantum mechanics, I mention what Louis de Broglie wrote in 1963: 'I suddenly had the idea, during the year 1923, that the discovery made by Einstein in 1905 should be generalized by extending it to all material particles.'[16] In other words, not only light but also matter should exhibit dual particle–wave behavior, a daring idea amply confirmed later.

As the preceding account shows, Einstein was not only one of the three fathers of the old quantum theory, but also the godfather of wave mechanics. Yet he could never accept the fusion of particles and waves, which he had anticipated in 1909, after it finally arrived in 1925. I have often wondered why but have no good answer. It seems to me that this belongs not just to his intellectual but also to his emotional make-up. His other work shows such flexibility, such daring, yet in regard to quantum mechanics he revealed such rigidity. I may add that to the end of his life he thought much and hard about quantum physics. He did not just say: I don't like what is happening. He searched continually for alternatives.

Einstein is the only scientist to be justly held equal to Newton. That comparison is based exclusively on what he did before 1925. In the remaining 30 years of his life he remained active in research but his fame would be undiminished, if not enhanced, had he gone fishing instead. Which completes my explanation of what Heisenberg wrote in 1963.

Bohr entered the arena of quantum mechanics in the autumn of 1927, shortly after Heisenberg had published his uncertainty relations. Bohr focused then and for the rest of his life on the language of science, the way in which we communicate. Straight away in 1927 he stated his main theme: 'Our interpretation of the experimental material rests essentially upon the classical concepts.'[17] It sounds simple enough, but is also most profound. Let me enlarge. In the classical era one verified the validity of theories by comparing them with experimental observations made with balances, thermometers, voltmeters, etc. The theories have been modified in the quantum era but – and that was Bohr's point – their validity continues to be verified by the same readings of the equilibrium position of a balance, the mercury column of a thermometer, the needle of a voltmeter, etc. The phenomena may be novel, their modes of detection may have been modernized, but detectors should be treated as classical objects; their readings continue to be described in classical terms.

'The situation thus created is of a peculiar nature,' Bohr remarked.[17] Consider for example the question: Can I not ask for quantum-mechanical properties of a detector, say a voltmeter? The answer is yes, I can. Next question: but should I then not abandon the limited description of the voltmeter as a classical object, and rather treat it quantum-mechanically? The answer is yes, I must. But in order to register the quantum properties of the voltmeter I need *another* piece of apparatus with which I again make classical readings.

On another issue of language Bohr said (I paraphrase): 'The question: *is* an electron a particle or *is* it a wave? is a sensible question in the classical context where the relation between object of study and detector either needs no specification or else is a controllable relation. In quantum mechanics that question is meaningless, however. There one should rather ask: does the electron (or any other object) *behave* like a particle or like a wave? That question is answerable, but only if one specifies the experimental arrangement by means of which "one looks" at the electron.

To summarize, Bohr stressed that only by insisting on the description of observations in classical terms can one avoid the logical paradoxes apparently posed by the duality of particles and waves, two terms themselves defined classically. Wave and particle behavior mutually exclude each other. The classical physicist would say: if two descriptions are mutually exclusive then at least one of them must be wrong. The quantum physicist will say: whether an object behaves as a particle or as a wave depends on your choice of experimental arrangement for looking at it. He will not deny that particle and wave behavior are mutually exclusive but will assert that both are necessary for the full understanding of the properties of the object. Bohr coined the term *complementarity* for describing this new situation:

> The very nature of the quantum theory forces us to regard particle behavior and wave behavior, the union of which characterized the classical theories, as complementary but exclusive features of the description. Complementary pictures of the phenomena only together offer a natural generalization of the classical mode of description.

In the course of time Bohr improved and refined the language of observation, culminating in 1948 in his precise definition of the concept 'phenomenon':

> Phrases often found in the physical literature, as 'disturbance of phenomena by observation' or 'creation of physical attributes of objects by measurements' represent a use of words like 'phenomena' and 'observation' as well as 'attribute' and 'measurement' which is hardly compatible with common usage and practical definition and, therefore, is apt to cause confusion. As a more appropriate way of expression, one may strongly advocate limitation of the use of the word *phenomenon* to refer exclusively to observations obtained under specified circumstances, including an account of the whole experiment.[18]

Bohr's usage of 'phenomenon' is the one by now subscribed to by nearly all physicists.

Not by Einstein, however. Until his death he kept insisting that one should look for a deeper level of theory in which one can talk of phenomena independently of observational details. He did eventually accept the results of quantum mechanics but believed that these should emerge by some effective averaging process applied to the deeper theory. Bohr tried hard, and often, to convince him of the complementarity view. He never succeeded.

To conclude I return to questions raised at the beginning.

What did Bohr really do? He was not only a major figure in physics but also one of the most important twentieth-century philosophers. As such he must be considered the successor to Kant, who had considered causality as a 'synthetic judgement a priori', not derivable from experience. Causality is, in Kant's own words, 'a rule according to which phenomena are sequentially determined. Only by assuming that rule is it possible to speak of experience of something that happens.' This view must now be considered *passé*. Since Bohr the very definition of what constitutes a phenomenon has undergone changes that, unfortunately, have not yet sunk in sufficiently among professional philosophers.

Again according to Kant, constructive concepts are intrinsic attributes of the '*Ding an sich*', a viewpoint desperately maintained by Einstein, but abandoned by quantum physicists. In Bohr's words: 'Our task is not to penetrate in the essence of things, the meaning of which we don't know anyway, but rather to develop concepts which allow us to talk in a productive way about phenomena in nature.'[19] After Bohr's death Heisenberg wrote that Bohr was 'primarily a philosopher, not a physicist',[3] a judgement that is arguable yet particularly significant if one recalls how greatly Heisenberg admired Bohr's physics.

My answer to the question raised in the 1980s by my friend is now complete.

Why do some textbooks not mention complementarity? Because it will not help in quantum-mechanical calculations or in setting up experiments. Bohr's considerations are extremely relevant, however, for the scientist who occasionally likes to reflect on the meaning of what she or he is doing.

Do we now possess the last word on quantum mechanics? I have seen too much to believe that anyone has the last word on any scientific issue, but I do think that Bohr's exegesis of the quantum theory is the best one we have to date.

References

1. See W. Schröder, *Nachr. Akad. der Wiss. Göttingen, math.-phys. Klasse*, 1985, p. 85.

2. P. Bridgman, letter to J.C. Slater's father, 4 February 1924; copy in the Library of the American Philosophical Society, Philadelphia.

3. W. Heisenberg, *Jahrb. der Bayer. Akad. der Wiss.*, 1963, p. 204.

4. E.N. da C. Andrade, *Proc. Roy. Soc.* A, **244**, p. 437, 1958.

5. A. Einstein in *Albert Einstein, philosopher–scientist*, ed. P.A. Schilpp, Tudor, New York, 1949.

6. A. Einstein, letter to B. Becker, 20 March 1954.

7. E. Bergner, *Bewundert und viel gescholten*, p. 212, Bertelsman, Munich, 1978.

8. N. Bohr, interview by T.S. Kuhn *et al.*, 17 November 1962; copy in Niels Bohr Archive, Copenhagen.

9. A. Einstein, letter to N. Bohr, 2 May 1920.

10. N. Bohr, letter to A. Einstein, 2 May 1920.

11. A. Einstein, letter to H.A. Lorentz, 4 August 1920.

12. N. Bohr, letter to A. Einstein, 11 November 1922.

13. A. Einstein, letter to N. Bohr, 11 January 1923.

14. N. Bohr, interview by A. Bohr and L. Rosenfeld, 12 July 1961.

15. A. Einstein, *Phys. Zeitschr.*, **10**, 185, 817, 1909.

16. L. de Broglie, preface to his re-edited Ph.D. thesis, 'Recherches sur la theorie des quanta', p. 4, Masson, Paris, 1963.

17. N. Bohr, *Nature*, **121** (suppl.), 580, 1928.

18. N. Bohr, *Dialectica*, **2**, 312, 1978.

19. N. Bohr, letter to H.P.E. Hansen, 20 July 1955.

3

*De Broglie, Einstein, and the birth of the matter wave concept**

I never had the privilege of meeting Louis de Broglie in person. I did, however, at one time have an exchange of letters with him which, I believe, is of sufficient interest to be recorded in this collection of *témoignages* to an important figure in the development of the quantum theory.

This correspondence took place in the summer of 1978 while I was in the process of writing Einstein's scientific biography.[1] In order to explain what was at issue it is necessary to recall a few dates. First, de Broglie's communication of 10 September 1923, in which 'After long reflection and meditation I suddenly had the idea . . . that the discovery [of the photon] by Einstein in 1905 should be generalized by extending it to all material particles'.[2] Accordingly he proposed[3] that the relation $E=h\nu$ between energy E and frequency ν shall hold not only for photons but also for electrons, to which he assigned 'a fictitious associated wave'. Then followed a second paper[4] (24 September) in which he indicated the direction in which one 'should seek experimental verifications of our ideas': a stream of electrons traversing an aperture whose dimensions are small with the electron wavelength 'should show diffraction phenomena'. These papers mark the birth of particle–wave duality applied to matter. One month later de Broglie presented a third short paper[5] which I find particularly interesting because it contains the first derivation of the density of states of electromagnetic radiation by means of a counting procedure in the phase space of photons – one year before Satyendra Nath Bose would independently do the same in his derivation of Planck's radiation law.

These three papers were extended to form de Broglie's doctoral thesis, which he defended on 25 November 1924. Three weeks later Einstein

* Reprinted with permission from *Louis de Broglie que nous avons connu*, published by Fondation Louis de Broglie, Paris, 1988.

wrote[6] to Lorentz: 'A younger brother of [Maurice] de Broglie has undertaken a very interesting attempt to interpret the Bohr–Sommerfield quantum rules (Paris dissertation 1924). I believe it is the first feeble ray of light on this worst of our physics enigmas. I, too, have found something which speaks for this construction.' Thus Einstein's response came rapidly. The subject of my correspondence with de Broglie was: When did Einstein first know of de Broglie's work and how did he know it?

There was a more specific reason for my query. From 21 to 27 September 1924 a physics meeting was held at Innsbruck. Pauli has written about this meeting: 'The author [P.] remembers that, in a discussion at [this] physics meeting in the autumn of 1924, Einstein proposed the search for interference and diffraction phenomena with molecular beams.'[7] That was two months before de Broglie's thesis defense. Had Einstein read the de Broglie papers in the *Comptes Rendus*? Or had he theoretically deduced the wave properties of matter independently? That was not out of the question because of the 'something' which Einstein had found, and to which I shall come back. I decided to pose my questions to Professor de Broglie. Here, in translation, in his response.

> 9 August 1978
> Dear Sir,
>
> I have received your very interesting letter and I shall try to answer your questions.
> 1. I do not believe that Einstein knew the three Notes that I had published in the *Comptes Rendus de l'Académie des Sciences* in 1923, Notes in which I sketched the ideas which I developed subsequently in my Thesis. Nevertheless, since Einstein would receive the *Comptes Rendus* and since he knew French very well, he might have seen my Notes.
> 2. When in 1923 I wrote the text of the Doctor's Thesis, which I wanted to submit in order to obtain the *Doctorat ès Sciences* I had *three* typed copies made and I handed one of these to M. Langevin so that he could decide whether the text could be accepted as a Thesis. M. Langevin, probably a bit astonished by the novelty of my ideas, asked me for a *second* typed copy of my Thesis, in order to send it to Einstein. It was then that Einstein, after reading my work, declared that my ideas seemed quite interesting to him, which made Langevin decide to accept my work . . .

On 18 August 1978 I wrote to de Broglie again, thanking him for his letter which had made evident that Einstein had seen the thesis before its defense. I further worte: 'Permit me one final question: Would you be able to recall *when precisely* Langevin sent a copy to Einstein?' de Broglie's reply follows.

28 September 1978

Dear Sir,

> I have received your letter and believe that I can respond with precision.
> I submitted a typed copy of my Thesis to M. Langevin in the begining of 1924 so that he could examine it. M. Langevin, eager to have the opinion of Einstein, asked me for a second copy of my Thesis which I was able to give him. I am therefore certain that Einstein knew of my Thesis *since the spring of 1924*. The necessary formalities for the acceptance and publication of a Thesis did not allow me to defend it until November 1924, but I believe I can affirm that by then Einstein knew the text of my Thesis since at least six months.

I was now able to write what became Chapter 24 of my Einstein biography.[1] I sent a draft copy to de Broglie for his approval and received the following reply:

12 January 1979

Dear Sir

> I have received your very interesting letter and your quite exact exposé of the origins of wave mechanics. I quite agree with the exposé you have given of the influence of Einstein on the development of my thinking at the time of my first papers . . .

To conclude I return to the 'something' that Einstein had found. Stimulated by a paper by S.N. Bose, Einstein had spent much of the second half of 1924 working on the quantum properties of a molecular gas. The most widely known result of these labors, which resulted in three papers, is the phenomenon that came to be known as Bose–Einstein condensation. The most profound result is contained in the second paper[8] submitted on 8 January 1925. It deals with fluctuations in the number of molecules in some given energy interval. His formula for this quantity consists of two terms.[9] One is the familiar Poisson distribution. The other was unfamiliar for particles but looked like a term that Einstein had encountered years earlier in his studies of energy fluctuations for radiation. It looked in fact like a fluctuation term for wave phenomena.

Hence Einstein was led to 'interpet [the second term] in a corresponding way for the gas, by associating with the gas a radiative phenomenon . . . I pursue this interpretation further, since here we have to do with more than a mere analogy.'[8] And then he adds: 'How a wave field can be associated with a material particle or a system of material particles has been explicated by Hr. de Broglie in a very noteworthy document.' A footnote referring to de Broglie's thesis is added.[9]

Thus did Einstein acknowledge the value of de Broglie's idea not only right before but also immediately after the appearance of the thesis.

The only personal encounters between the two men took place in the autumn of 1927, in Brussels and Paris. Twenty-five years later de Broglie wrote: 'The paper [of January 1925] by M. Einstein drew attention to my work which till then had been little noticed. For this reason I have always felt much personal gratitude towards him for the precious encouragement which he thus brought me'.[10]

Facsimile of letters by the 86-year old Louis de Broglie to the Author, dealing with events surrounding the publication of de Broglie's doctoral thesis. (Author's private collection.)

26 Septembre 1978

Cher Monsieur

J'ai bien reçu votre lettre et je crois pouvoir y répondre avec exactitude.

J'ai remis un exemplaire dactylographié de ma Thèse dès le début de 1924 à M. Langevin pour qu'il l'examine. M. Langevin, désireux d'avoir l'avis d'Einstein sur ce travail, m'a demandé un deuxième exemplaire de ma Thèse que j'ai pu lui donner. Je suis donc certain qu'Einstein a eu connaissance de ma Thèse dès le printemps de 1924. Les formalités nécessaires pour l'acceptation et la publication de ma Thèse ne m'ont permis de la soutenir qu'en Novembre 1924, mais je crois pouvoir affirmer qu'Einstein connaissait alors le texte de ma Thèse depuis au moins six mois.

Veuillez agréer, cher Monsieur, l'expression de mes sentiments les meilleurs

Louis de Broglie

9 Août 1978

Cher Monsieur

J'ai bien reçu votre très intéressante lettre et je vais essayer de répondre à vos questions.

1° Je ne crois pas qu'Einstein avait eu connaissance des trois Notes que j'avais publiées dans les Comptes Rendus de l'Académie des Sciences en 1923, notes où j'esquissais les idées que j'ai ensuite développées dans ma Thèse. Néanmoins, comme Einstein devait recevoir les Comptes Rendus et qu'il savait très bien le français, il pouvait avoir remarqué mes Notes.

2° Quand j'ai écrit, en 1923, le texte de la Thèse de Doctorat que je voulais présenter pour obtenir le Doctorat ès Sciences, j'ai fait faire trois exemplaires dactylographiés et j'ai remis un de ces exemplaires à M. Langevin pour qu'il décide si ce texte pouvait être accepté comme Thèse. M. Langevin, probablement un peu étonné par la nouveauté de mes idées, m'a demandé de lui fournir un deuxième exemplaire dactylographié de ma Thèse pour l'envoyer à Einstein. C'est alors qu'Einstein, après lecture de mon travail, a déclaré que mes idées lui paraissaient tout à fait intéressantes, ce qui a décidé Langevin à accepter mon travail.

Je me permets de vous signaler que mon principal collaborateur actuel est M. Georges Lochak dont voici l'adresse: 67 Boulevard de Picpus 75012 Paris. Vous pourriez entrer en relations avec lui.

Veuillez agréer, cher Monsieur, l'expression de mes sentiments les meilleurs

Louis de Broglie

12 janvier 1979

Cher Monsieur

J'ai bien reçu votre très intéressante lettre et votre très exact exposé des origines de la Mécanique Ondulatoire. Je suis bien d'accord avec l'exposé que vous faites de l'influence d'Einstein sur le développement de ma pensée au moment de mes premiers travaux. Je vous signale qu'à la page 63, il vaut mieux remplacer "to the French Academy" par "to the French Academy of Sciences", car en France on appelle "Académie française" l'Académie des Lettres. Mais ce n'est là qu'un détail.

Je me permets de vous envoyer ci-joint un exemplaire de mon livre "Nouvelles perspectives en Microphysique" où vous trouverez aux pages 125 et 260 des articles qui pourront vous intéresser.

Je vous adresse également un exemplaire de mon livre "Recherches d'un demi-siècle" qui contient un grand nombre d'intéressants exposés.

Je suis à votre disposition pour vous fournir d'autres renseignements qui pourraient vous être utiles et je vous prie de croire, cher Monsieur, à mes sentiments dévoués

Louis de Broglie

References

1. A. Pais, *Subtle is the Lord*, Clarendon Press, Oxford, 1982.

2. L. de Broglie, preface to his reedited 1924 Ph.D. thesis, p. 4. Masson, Paris, 1963.

3. L. de Broglie, *C. R. Acad. Sci. Paris*, **177**, 507, 1923.

4. L. de Broglie, *C. R. Acad. Sci. Paris*, **177**, 548, 1923.

5. L. de Broglie, *C. R. Acad. Sci. Paris*, **177**, 630, 1923.

6. A. Einstein, letter to H.A. Lorentz, 16 December 1924.

7. W. Pauli in *Albert Einstein: Philosopher-Scientist* (ed. P.A. Schilpp), p. 156. Tudor, New York, 1949.

8. A. Einstein, *Sitz. Ber. Preuss. Ak. Wiss.* 1925, p. 3.

9. For more details see A. Pais, *Subtle is the Lord*, Clarendon Press, Oxford, 1982, Chapter 24, section (b).

10. L. de Broglie, *Nouvelles perspectives en microphysique*, p. 233, Albin Michel, Paris, 1956.

4

*Einstein, Newton, and success**

If I had to characterize Einstein by one single word I would choose *apartness*. This was forever one of his deepest emotional needs. It was to serve him in his single-minded and single-handed pursuits, most notably on his road to triumph from the special to the general theory of relativity. It was also to become a practical necessity to him, in order to protect his cherished privacy from a world hungry for legend and charisma. In all of Einstein's scientific career apartness was never more pronounced, however, than in regard to the quantum theory. This covers two disparate periods. From 1905 to 1923 he was the only one, or almost the only one, to take seriously his own light-quantum hypothesis: under certain circumstances light behaves as if it has a particulate structure. During the second period, from 1926 to the end of his life, he was the only one, or again nearly the only one, to maintain a profoundly sceptical attitude to quantum mechanics.

Yet Einstein has called 'the statistical quantum theory [i.e. quantum mechanics] . . . the most successful physical theory of our period'.[1] Then why was he not convinced by it? I believe Einstein himself answered this indirectly in his 1933 Spencer lecture, *On the Method of Theoretical Physics* – perhaps the clearest and most revealing expression of his mode of thinking. The key is to be found in his remarks on Newton and classical mechanics.

In this lecture Einstein notes that 'Newton felt by no means comfortable about the concept of absolute space, . . . of absolute rest . . . [and] about the introduction of action at a distance'. Einstein then goes on to refer to the success of Newton's theory in these words: 'The enormous practical

* Reprinted with permission from *Einstein, a centenary volume*, ed. A.P. French, Harvard University Press, 1979.

success of his theory may well have prevented him and the physicists of the eighteenth and nineteenth centuries from recognizing the fictitious character of the principles of his system.' (It is important to note that by 'fictitious', Einstein means free inventions of the human mind.) Whereupon he compares Newton's mechanics with his own work on general relativity: 'The fictitious character of the principles is made quite obvious by the fact that it is possible to exhibit two essentially different bases each of which in its consequences leads to a large measure of agreement with experience.'[2]

Now back to the quantum theory. In the Spencer lecture Einstein mentioned the success not only of classical mechanics but also of the statistical interpretation of quantum theory. 'This conception is logically unexceptionable and has led to important successes.' But, he added, 'I still believe in the possibility of giving a model of reality which shall represent events themselves and not merely the probability of their occurrence.'[2]

From this lecture as well as from numerous discussions with him on the foundations of quantum physics I have gained the following impression. Einstein tended to compare the successes of classical mechanics with those of quantum mechanics. In his view both these theories were on a par, being successful but incomplete. For more than a decade Einstein had pondered the single question of how to extend to general motions the invariance under uniform translations. His resulting theory of 1916, general relativity, had led to only small deviations from Newton's theory. (Large deviations were discussed only much later.) He was likewise prepared to undertake the search for his own 'model of reality', no matter how long it would take, and he was also prepared for the survival of the practical successes of quantum mechanics, with perhaps only small modifications. It is quite plausible that the very success of his highest achievement, general relativity, was an added spur to Einstein's apartness. Yet it should not be forgotten that this trait characterizes his entire *oeuvre* and mode of life.

What did Einstein want? It is essential for the understanding of his thinking to realize that there were two sides to his attitude concerning quantum mechanics. There was Einstein the critic, never yielding in his dissent from complementarity, according to which the notion of 'physical phenomenon' *irrevocably* includes the specifics of the experimental conditions of observation. And there was Einstein the visionary, ever trying to realize an 'objectively real' world model, a deeper-lying theoretical framework that permits the description of phenomena independently of these conditions. He believed that quantum mechanics should be deducible as a limiting case of such a future theory 'just as electrostatics is deducible from the Maxwell equations of the electromagnetic field or as thermodynamics is deducible from statistical mechanics'. He did not believe that quantum mechanics itself was a useful starting-point in the search for this future theory, 'just as one cannot arrive at the foundations of mechanics from thermodynamics or statistical mechanics'.[3]

This vision that Einstein pursued can be traced back at least to 1920, well before the advent of quantum mechanics. It was a unified field theory. But by that he meant something different from what it meant and means to anyone else. He demanded that it be a local field theory, causal in the classical sense; that it unify the forces of nature; that the particles of physics shall emerge as special solutions of the general field equations; and that the quantum postulates *should be a consequence* of these equations.

Einstein was neither saintly nor humourless in defending his solitary position on quantum mechanics, nor was he oblivious to the negative reaction to it by others. He may not have expressed all his feelings on these matters. But that was his way. 'The essential of the being of a man of my type lies precisely in *what* he thinks and *how* he thinks, not what he does or suffers.' In any event he held fast. 'Momentary success carries more power of conviction for most people than reflections on principle.'

Yet as his life drew to a close, occasional doubts on his vision arose in his mind. In the early fifties he once said to me, in essence: 'I am not sure that differential geometry is the framework for further progress, but if it is, then I believe I am on the right track'. Similar reservations are also found in his letters of that period to Max Born and to his lifelong friend Michele Besso.[4]

Otto Stern has recalled a statement that Einstein once made to him: 'I have thought a hundred times as much about the quantum problems as I have about general relativity theory.'[5] Einstein kept thinking about the quantum till the very end. He wrote his last autobiographical sketch in Princeton, in March 1955, about a month before his death. Its final sentences deal with the quantum theory. 'It appears dubious whether a field theory can account for the atomistic structure of matter and radiation as well as of quantum phenomena. Most physicists will reply with a convinced "No", since they believe that the quantum problem has been solved in principle by other means. However that may be, Lessing's comforting words stay with us: "The aspiration to truth is more precious than its assured possession."'[6]

References

1. A. Einstein, in *Albert Einstein: philosopher–scientist*, Library of Living Philosophers, 1949; Evanston, p. 1.

2. A. Einstein, *On the method of theoretical physics*, Oxford University Press, 1933; reprinted in *Phil. Sci.* **1**, 162, 1934.

3. A. Einstein, *J. Franklin Inst.* **221**, 313, 1936.

4. A. Einstein, letter to M. Besso, 24 July 1949.

5. R. Jost, letter to A. Pais, 17 August 1977.

6. A. Einstein, in *Helle Zeit, Dunkle Zeit*, ed. C. Seelig, Europa Verlag, Zürich, 1956.

5

*A mini-briefing on relativity, for the layman**

Einstein's theory of the Brownian motion, his introduction of the light quantum, but above all his theories of special and general relativity have profoundly changed the way modern men and women think about the phenomena of inanimate nature. It would actually be better to say 'modern scientists' than 'modern men and women'. Indeed, in order to appreciate the contributions by Einstein in their fullness one needs to be schooled in the physicist's style of thinking and in mathematical techniques – not that difficult for special relativity but quite advanced for the case of general relativity. It is an optimistic guess that only one in every hundred thousand people alive today has any insight into what Einstein's relativity is really all about.

It is of course by no means uncommon that only a tiny fraction of humanity has a genuine grasp of advances in science of any kind whatsoever, or even knows who made them. It is therefore all the more remarkable and significant that the name Einstein is so familiar to so very many. It is no exaggeration to say that he was one of the rare mythical figures in his lifetime and that he is still so today. What is even more extraordinary is that he became a world figure precisely because of his most abstruse work, his theory of general relativity. Thus in discussing the impact of relativity theory one needs to address two only marginally related questions: How did relativity affect scientific thinking? And how did it come about that Einstein's name became a byword in almost every culture?

Elsewhere in this book I have attempted to answer the second question. (See 'Einstein and the Press', pp. 148–150. Here I try to answer the first one in simple terms.

* Excerpt from my article 'Knowledge and belief', which appeared in *American Scientist*, Vol. 76, p. 154, 1988. Reprinted with permission.

Although it would be out of place to discuss the technical aspects of relativity theory here, I should briefly state its principal concepts. First, what is meant by special and what by general relativity? In the special theory one compares the experiences of observers that move uniformly, that is, in straight lines and with constant velocity, relative to each other. In the general theory all possible relative motions are considered, including curvilinear paths and nonconstant velocities.

To begin with, let us confine ourselves to the special theory – worked out by Einstein in 1905 – which is based on two fundamental postulates or axioms. (1) The laws of physics take the same form for any two observers that move relatively to each other in a straight line and with constant (time-independent) velocity (uniform motion). (2) Any observer will find the velocity of light to be the same whether the light be emitted by a body at rest or a body in uniform motion. Observe that neither postulate contains any statement concerning *what* the laws of nature are. Rather, these axioms refer exclusively to motions and are supposed to apply *whatever* these laws may be. In the language of physics they are kinematical, not dynamical, axioms.

Postulate 2 is in flagrant contradiction with everyday experience. If an observer standing on a platform sees a train going by with a speed of 100 km/hour, and further sees a man walking forward in the train with a speed of 5 km/hour relative to the train, then this observer will see the man moving forward with a speed of 105 km/hour. But if the walking man is replaced by a light beam, then the observer on the platform will find the same value for the light velocity whether the train stands still or is in motion!

Here we encounter the first scientific impact of relativity: what is observed in everyday life experience cannot necessarily be extrapolated to all experience. Special relativity has taught us to be extremely careful in attempting to extrapolate to things moving nearly as fast as light. General relativity has taught us that, in some respects, care is also called for in extrapolating to things that are as heavy as stars. Quantum mechanics (not part of this account) in turn calls for care in extrapolating to things that are as small as atoms. Those are the physicist's cautionary tales of the twentieth century.

The universal velocity of light is so large (about 300 000 km/second) that it is in fact not part of the observer's everyday experience. Did the advent of special relativity then leave all earlier everyday conclusions unaffected? As a point of principle the answer is 'no'. For instance, this theory tells us that the answer '105 km/hour' in the example of the train needs correction; the right answer is actually one ten-thousandth of a billionth of one per cent less. That modification is so incredibly small, however, that if not in principle then in practice the wisdom of our fathers continues to apply to trains or space vehicles. This also holds true for other consequences of special relativity that would have appeared as most bizarre to our ancestors, as in fact they did to some of Einstein's contemporaries.

To give another example: Observers A and B at rest with respect to each other synchronize watches and equalize measuring rods. When next B moves uniformly away from A, then A will observe B's watch to run behind his own and B's rod to be shortened compared to his own; while B comes to the same conclusion in respect to A's tools. Again, in practice these effects are observable only for relative velocities comparable to the velocity of light.

One final comment about modifications brought about by special relativity. In the nineteenth century, scientists had identified the conservation of energy and the conservation of mass as two separate strict natural laws. Not so. Relativity tells us that mass (m) is a form of energy (E): $E=mc^2$ (c is the velocity of light in a vacuum). That new view does not, however, affect the butcher when weighing meat. Nor will it affect us in any other respect in our daily lives – as long as we shall be spared a nuclear holocaust.

As to content, special relativity is exclusively based on the two postulates already mentioned. As to form, enormous simplifications were attained (in the first instance, due to the physicist Hermann Minkowski) by expressing the basic equations of physics in terms of a four-dimensional 'space–time geometry': three dimensions for space, one for time. It should be stressed that this most useful mathematical device does not affect the content of the theory beyond what has been sketched above. In particular, space remains flat, as had been tacitly assumed since the beginnings of theoretical physics.

After an intense intellectual struggle Einstein realized that the transition from special to general relativity could be achieved only by renouncing this flatness properly. In general relativity theory one includes all possible kinds of relative motion rather than just uniform motions. Postulate (1) is taken to apply also in this more general situation. This demands that space is curved, he now asserted, the amount of curvature at any particular place depending upon how dense matter is at that place. One may roughly compare this state of affairs with that of a man jumping on a trampoline, which is flat after he has jumped up but which curves when he hits the surface. So it is with space, or rather space–time, which since Newton had been considered as the flat empty stage on which matter moves and our destinies unfold. Since the advent of general relativity it has become clear, however, that matter by its gravitational action actually determines 'what shape space is in', like the jumping man on the trampoline. Far away from matter space remains flat and Euclidean, but where matter is present space obeys the more general (Riemannian) geometry.

Nor does gravitation any longer obey the laws laid down by Newton. Yet it would be wrong to say that Newton's theory of gravitation is wrong; rather it is approximate. The laws he deduced upon seeing the apple fall from the tree (actually a true story, though the apple did not hit him on the head) do indeed apply to the apple, not with ultimate mathematical rigor, but to an extraordinarily high approximation. As Einstein was the first to show, his refined gravitational equations reproduce all of Newton's results in the limit where gravitational forces are weak. To see deviations from Newton's laws one needs to consider gravitational attractions that are vastly stronger than those between a lightweight apple and the Earth, which is also a lightweight object from the point of view of general relativity. Once again, one must be careful in extrapolating from everyday experience.

The superiority of Einstein's over Newton's theory first became manifest in the motion of the planet closest to the sun and most strongly affected by the sun's gravitational pull – Mercury. It had been known since 1859 that Mercury's orbit shows a tiny anomaly. Newton's laws predicted its perihelion to be stationary, yet actually it was observed to advance a little, about 43 seconds per century (after corrections for disturbances from nearby planets). Decades of effort at ascribing this effect to a perturbing influence (another as yet unobserved tiny planet, for example), had proved fruitless. So it remained until late 1915, when Einstein was able to show quantitatively that this anomaly was a necessary consequence of his new theory. From that time on he knew: Nature had spoken; he had to be right.

While this was a monumental discovery, it did not generate widespread attention to its author. That came about only a few years later, in 1919, when another prediction of his theory was verified: the bending of light. I have described in Chapter 11 of this book, 'Einstein and the Press' (sections 3 and 4), how the events of that year brought about Einstein's lasting world fame.

Here my briefing ends. Want to know more about Einstein and relativity? *Subtle is the Lord*, my biography of him, may serve you as a further guide.

6

*How Einstein got the Nobel Prize**

On 10 November 1922, a telegram was delivered to the Einstein residence in Berlin. It read: 'Nobelpreis für Physik ihnen zuerkannt näheres schriftlich [signed] Aurivillius' ('Nobel Prize for Physics awarded to you; more by letter'). On the same day, a telegram with the identical text must have been received by Niels Bohr in Copenhagen. Also on that day, Professor Christopher Aurivillius, secretary of the Royal Swedish Academy of Sciences, wrote to Einstein: 'As I have already informed you by telegram, in its meeting held yesterday, the Royal Academy of Sciences decided to award you last year's Nobel Prize for Physics, in consideration of your work in theoretical physics, and in particular your discovery of the law of the photoelectric effect, but without taking into account the value which will be accorded your theories of relativity and gravitation, once they are confirmed.'[1] On the day that Einstein was awarded the 1921 Prize, the 1922 Prize went to Bohr 'for his services in the investigation of the structure of atoms and of the radiation emanating from them'.[2]

Almost exactly three years earlier, on 6 November 1919, the Royal Society and the Royal Astronomical Society had convened a joint session in London, at which Crommelin and Eddington had presented evidence for the bending of light, obtained during observations of the total solar eclipse of 29 May 1919.[3] Before departing for his observing station on the Island of Principe, Eddington had written: 'The present eclipse expeditions may for the first time demonstrate the weight of light; or they may confirm Einstein's weird theory of non-Euclidean space; or they may lead to a result of yet more far-reaching consequences – no deflection.'[4]

It was a momentous issue. 'Do not Bodies act upon Light at a distance,

* Reprinted from *American Scientist*, Vol. 70, No. 4, July–August, 1982, pp. 358–65.

and by their action bend its Rays; and is not this action (*caeteris paribus*) strongest at the least distance?' Newton[5] had asked in 1704 in Query 1 of his *Opticks*. In 1911, when Einstein had not yet realized that space is curved, he had computed the answer to Newton's Query, namely, that light coming from a distant star is bent by 0.87 seconds of arc as it grazes the Sun on its way to the Earth.[6] (He was not the first to obtain this result, but that is another story.) Eddington referred to this so-called Newtonian value when he wrote of 'the weight of light.' In 1915, Einstein recalculated this effect on the basis of his 'weird theory,' the general theory of relativity, and this time, he came up with a value of 1.74 seconds of arc, twice the Newtonian value.[7] The eclipse expedition had set out to decide between these two values.

On 6 November 1919, Crommelin and Eddington had reported that the observations confirmed Einstein's value.[3] On that occasion, Sir Frank Dyson, the Astronomer Royal, stated: 'After a careful study of the plates, I am prepared to say that they confirm Einstein's prediction . . . Light is deflected in accordance with Einstein's law of gravitation.' J.J. Thomson, who was in the Chair, remarked: 'This is the most important result obtained in connection with the theory of gravitation since Newton's day . . . The result [is] one of the highest achievements of human thought.' In 1919, the Dutch theoretical physicist Lorentz, a man not easily given to hyperbole, wrote to Ehrenfest, his successor as professor at Leiden, that the result of the eclipse expedition was 'one of the most brilliant confirmations of a theory ever achieved [and is] also very suitable for paving the way to the Nobel Prize.'[8] Thus, men of great eminence expressed themselves in the strongest terms about the importance of the new results on the bending of light.

Moreover, in 1915, on the basis of the theory of general relativity, Einstein had been able to predict a value for the precession of the perihelion of the planet Mercury – a small effect that had remained unexplained for more than half a century – that is in splendid agreement with the observed value. Then why should Aurivillius write to Einstein that the Swedish Academy of Sciences made its decision without taking into account the value of either the special or general theory of relativity? Why, the question has often been asked, did Einstein not get the Prize for relativity?

The question of how Einstein did in fact get the Nobel Prize will be discussed in this essay on the basis of letters nominating him for this award and of reports of the deliberations of the Nobel Committee for Physics. The Academy's decisions have nearly always been well received by the community of physicists. To be sure, eyebrows (including my own) are raised on occasion. That, however, is not only inevitable but also irrelevant to the account about to be given. My sole focus will be on matters of great historical interest: the scientific judgments of the leading scientists who made the nominations and of a highly responsible, rather conservative body of great prestige, the Nobel Committee. The story has neither heroes nor culprits.

Einstein: German or Swiss?

Einstein was not home to receive Aurivillius's telegram and letter. He and his second wife, Elsa, were on their way to Japan. Shortly before their departure, Laue, professor of theoretical physics in Berlin, had written to him: 'According to information I received yesterday, which is certain, events may occur in November that might make it desirable for you to be present in Europe in December. Consider whether you will nevertheless go to Japan.'[9] Einstein left anyway and would not be back in Berlin until March 1923.

The previous three years had been a hectic period in his life. In January 1919, he and Mileva, his first wife, were divorced. At that time, he had promised her the money he would receive when his Nobel Prize came. In 1923, all of his 121,572 kronor and 54 öre (about $32,000) were indeed transmitted to her. In June 1919, he married his cousin Elsa. In November, the publicity following the news about the bending of light suddenly turned him into a charismatic world figure. The next year was a difficult one. In January, his mother, mortally ill, came to Berlin to spend her final days with her beloved son and died shortly afterward in his home. Later that year, Einstein's integrity and work both came under attack from some quarters in Germany. The following year, he traveled to the United States and England. In June 1922, during the violence that accompanied the uncertain rise of the Weimar Republic, Germany's foreign minister, Walther Rathenau, a Jew and a good acquaintance of Einstein's, was murdered.

Rumors circulated that Einstein's life, too, might be in danger. When he set out for Japan, he was glad to absent himself for a while from a potentially unsafe situation. The news of the Nobel award must have reached him while he was en route. I do not know, however, when and where he received word. In the travel diary he kept during the journey, I found no mention of this event. The Einsteins stayed in Japan from 17 November until 29 December 1922.

Meanwhile, on 10 December, Rudolf Nadolny, the German ambassador to Sweden, accepted the Nobel Prize in Einstein's name and, in a toast offered at the banquet held in Stockholm that evening, expressed 'the joy of my people that once again one of them has been able to achieve something for all mankind.' To this, he added 'the hope that Switzerland, which during many years provided the scholar with a home and opportunities to work, would also participate in this joy'.[10] In point of fact, Switzerland had provided Einstein with more than a home. In 1922, he was a Swiss citizen. How a Swiss citizen came to be represented by a German ambassador is a rather hilarious story.

Einstein was born in 1879 in Ulm in the Kingdom of Württemberg, which in 1871 had joined the newly founded state of Germany. He was thus German by birth. On 28 January 1896, then a high school student in Aarau, Switzerland, he received a document he had applied for, certifying that he was no longer a citizen of Württemberg. He remained stateless until 21 February 1901, when he became a Swiss citizen. In April 1914, Einstein moved to Berlin, where he had accepted a research position under the aegis of the Prussian Academy of Sciences. He was told that, since the position made him a state official, he must become a German citizen. He insisted, however, on retaining his Swiss citizenship. What happened next is not entirely clear, but it is certain that Einstein traveled to Japan on a Swiss passport. Presumably, the German authorities did not press the issue in 1914.

By 1919, his citizenship had already become a matter of national pride. A letter he received during that year from Stumpf, a fellow member of the Prussian Academy, reads as follows: 'Cordial congratulations on the occasion of the great new success of your theory of gravitation. With all our hearts, we share the elation which must fill you and are proud of the fact that, after the military-political collapse, German science has been able to score such a victory.'[11]

Ambassador Nadolny's report to the Foreign Office in Berlin, dispatched on 12 December 1922, shows that he had coped conscientiously with this problem in international relations. In November, he had been asked by the Swedish Academy of Sciences to represent Einstein. Next, the Swiss ambassador had asked for clarification, since to his knowledge Einstein was a Swiss citizen. On 1 December, Nadolny had cabled the University of Berlin for information. On 4 December, he had received a telegram from the Prussian Academy, which read: '*Antwort: Einstein ist Reichsdeutscher.*' On 11 December, the Foreign Office had informed him that Einstein was Swiss. On 13 January 1923, the Prussian Academy had informed the Kultusministerium in Berlin that, on 4 May 1920, Einstein had taken the oath as a state official and that, since only Germans could be state officials, he was therefore German. The protocol of the Prussian Academy of 18 January quoted the legal opinion that Einstein was a German citizen, but that his Swiss citizenship was not thereby invalidated. On 15 February, the Prussian Academy had informed Einstein of this ruling. On 24 March, Einstein had told the Prussian Academy that he had made no change in citizenship status a condition for accepting the position in Berlin. On 19 June, he had called in person on Ministerialrat Rottenburg of the Foreign Office and had reiterated his position, noting that he traveled on a Swiss passport. A note on this visit, prepared by Einstein on 7 February 1924 for inclusion in the *Acta* of the Prussian Academy, reads in part: '[Rottenburg] was of the decided opinion that my appointment to the Akademie implies that I have acquired Prussian citizenship, since the opposite opinion cannot be maintained on the basis of the *Acta*. I have no objections to this view.'[12]

It would appear that Einstein, never awed by officialdom, was highly amused. In any event, prior to this, on 6 April 1923, his stepdaughter Ilse had written to the Nobel Foundation in Stockholm that Professor Einstein would appreciate it if the medal and diploma could be sent to him in Berlin, adding that, if this were to happen via diplomatic channels, 'the Swiss Embassy should be considered, since Professor Einstein is a Swiss citizen.'[13] The matter ended when Baron Ramel, the Swedish ambassador to Germany, called on Einstein in his home and handed him his insignia.

(On 1 October 1940, Einstein was sworn in as a citizen of the United States. For the rest of his life, he retained his status as a Swiss citizen as well, however.)

Procedures

The procedure of the Royal Swedish Academy of Sciences (hereinafter called the Academy) for awarding the Nobel Prize in physics is as follows. Replies to invitations for nominations, sent out by the Academy, are handed over to a five-member Nobel committee (hereinafter called the Committee) elected from the membership. The Committee studies the proposals and supporting material, draws up a protocol of its deliberations, and decides by majority vote on a recommendation to the Academy. The recommendation is transmitted in the form of a report (hereinafter called the Report) that summarizes the merits of the nominations and gives the reasons for the Committee's decision. The recommendation is voted on first by the Academy Klass (Section) of physics, which need not agree with the Committee's recommendation, then by the full Academy, not just the physicists. Neither this final, decisive vote nor the preceding discussions of the full Academy are recorded. The Academy's decision need not agree with the Klass recommendation. In 1908, for example, the Committee unanimously, though hesitantly, recommended Max Planck for his work on black-body radiation; the Klass vote was also in support of Planck; but the full Academy chose Gabriel Lippmann for his work on color photography.

The case of Planck, recently discussed in detail by Nagel,[14] bears on the controversial nature of the quantum theory in its early days. In 1900, Planck had not only proposed the correct formula for the spectral energy density of electromagnetic radiation in thermal equilibrium – that is, black-body radiation – but had also discovered a primitive version of the quantum theory in his attempt to justify his formula. By 1908, it was clear that Planck's energy distribution agreed extraordinarily well with experiment. Yet his was still a formula in search of a definitive theory. I will not go into detail but will merely point out that those were the days of the 'old' quantum theory, which was not a theory of first principle, but rather a set of highly inspired guesses. In the 1908 Report Ångström, then chairman of the Committee, was entirely justified in expressing apprehensions that 'the theoretical treatment of the radiation problem is an as yet unfinished chapter, in which new efforts to solve the problem are both necessary and constantly tried.'

The apprehensions remained until 1925, when quantum mechanics was discovered. By 1918, however, it was evident that quantum ideas were fundamental and successful. Planck's formula for black-body radiation, as well as Einstein's formula for the photoelectric effect, had been found to agree beautifully with experiment. There was also a burgeoning quantum theory of specific heats. And there was Bohr's formula for the Rydberg constant.

Nominations for Planck continued to arrive in Stockholm, among them the first nomination for the physics prize ever made by Einstein.[15] Early in 1919, the Committee decided that the time was ripe to recognize Planck; later that year, the Academy voted accordingly.

Some of the Academy's major predicaments in the early twentieth century are well illustrated by the long time it took to adjudge Planck's case. Could the Prize be given for a formula that was indubitably of the greatest importance, although it was still in need of a solid derivation? Conversely, in the case of Einstein, could one give the award for the theories of special and general relativity, which were indubitably based on first principles, although they were not – or so it seemed to those responsible for assigning the Prize – sufficiently supported by experiment?

Nominations and deliberations

The records of the Nobel Committee for Physics show that Einstein received nominations for the physics prize every year between 1910 and 1922 except for 1911 and 1915. The Committee often divides the nominees into categories, in order to facilitate identification of leading contenders. For each of the years in which Einstein was nominated, Table 1 gives the name of the category in which he was included, the other nominees in the

Table 1. *Contenders for the Nobel Prize for Physics in the years in which Einstein was nominated and summary of the recommendations for Einstein*

Year	Category in which Einstein was nominated	Other nominees in that category	Winner and field	Comments
1910	Investigations of a theoretical or mathematical-physical character	Gullstrand, Planck, Poincaré	Van der Waals, equation of state	Campaign in support of Poincaré
1912	Theoretical physics	Heaviside, Lorentz, Mach, Planck	Dalén, automatic regulators for lighthouses and buoys	Sharing with Lorentz proposed by Wien and by Schaefer
1913	Theoretical physics	Lorentz, Nernst, Planck	Onnes, low-temperature physics	Count Zeppelin, Wright brothers nominated in another category
1914	Work of a more speculative nature, theoretical physics	Eötvös, Mach, Planck	Laue, X-ray diffraction by crystals	
1916	Molecular physics	Debye, Knudsen, Lehmann, Nernst		1916 physics prize never awarded
1917	Investigations connected with Planck's extremely fruitful researches concerning the quantum hypothesis	Bohr, Debye, Nernst, Planck, Sommerfeld	Barkla, X-ray spectra of elements	Prize awarded in 1918
1918	Quantum physics	Bohr, Paschen, Planck, Sommerfeld	Planck, discovery of energy quanta	Prize awarded in 1919; sharing with Lorentz proposed by Wien and Laue
1919	Theoretical physics	Knudsen, Lehmann, Planck, Perrin, Svedberg, Gouy	Stark, Doppler effect in canal rays and splitting of spectral lines in electric fields	
1920	Mathematical physics	Bohr, Sommerfeld	Guillaume, precision measurements in thermometry	
1921	Mathematical physics	Bohr, Sommerfeld	Einstein, photoelectric effect	Prize awarded in 1922. Hadamard suggested Einstein or Perrin
1922			Bohr, atomic structure	

category, and the winner and Prize-winning work for that year.

In October 1909, Wilhelm Ostwald, winner of the 1909 chemistry prize, to whom Einstein had applied unsuccessfully in the spring of 1901 for an assistantship, nominated Einstein for the 1910 physics prize.[16] He was the first to nominate Einstein and the only one to do so in 1909. He repeated his nominations for the 1912 and 1913 awards.[17,18] In all three instances, it was Einstein's work on special relativity that motivated him. In 1909, Ostwald wrote that special relativity was the most far-reaching new concept since the discovery of the energy principle. In his second nomination of 1911, he stressed the fact that special relativity frees man from bonds many thousands of years old. On the third occasion, in 1912, he emphasized, in opposition to some others, that the issues were of a physical rather than a philosophical nature and likened Einstein's contributions to the work of Copernicus and Darwin.

In nominating Einstein for the 1912 award, Ostwald was joined by the physicists Pringsheim, Schaefer, and Wien; for the 1913 award by Wien and Naunyn, a German professor of medicine. All these nominations were for special relativity only, though Naunyn added a remark on the quantum theory. Pringsheim wrote: 'I believe that the Nobel Committee will rarely have the opportunity of awarding a Prize for works of similar significance.'[19]

Wien's two nominations were actually for a sharing of the Prize between Einstein and Lorentz, and Schaefer also proposed this sharing as an alternative to Einstein alone. In his second letter of nomination, Wien wrote: 'Concerning the new experiments on cathode rays and beta rays, I would not consider them to have decisive power of proof. The experiments are very subtle, and one cannot be sure whether all sources of error have been excluded.'[20] At issue was the verification of Einstein's relation between the rest mass, m, the energy, E, and the velocity, υ, of a free electron:

$$E = mc^2/\sqrt{(1 - \upsilon^2/c^2)}$$

As early as 1908, some experimentalists claimed confirmation of this relation. Doubts remained, however, as Wien's letter shows; these were not fully dispelled until about 1915. Thus one important confirmation of special relativity achieved noncontroversial status only after the 1912 nominations had been made. Sommerfeld's theory of the fine structure of spectral lines, which uses a relation that is essentially the same as the Einstein relation, also came later, in 1916. By then, the momentous new developments of general relativity had changed the situation drastically. It is also worth noting that the energy–mass equivalence $E = mc^2$ was not verified experimentally until the 1930s.

Before turning in more detail to the later nominations, I will first note the reactions of the Committee to the earlier ones. The Report for 1910 suggested that further experimental verification was needed before the principle of special relativity could be accepted, and, in particular, before it could be awarded a Nobel Prize. The Report implied that the need for further confirmation was the reason why it was not until 1910 that Einstein was proposed, even though the principle in question had been put forward in 1905 and 'had caused the liveliest stir.' The Committee also noted that Einstein's work on Brownian motion had gained him great recognition.

The comments on special relativity in the Report for 1912 were in a similar vein, and it was noted that Lorentz, with whom Wien and Schaefer proposed that Einstein share the Prize, was more cautious with his hypotheses than Einstein. The Report for 1913 included a remark to the effect that special relativity was on its way to becoming a serious candidate for the award, even though the Committee expressed considerable doubt about likening Einstein to Copernicus or Darwin.

During the next few years, there was an inevitable lull. Einstein was deeply immersed in the struggle with general relativity and was confusing everybody, including himself, with a hybrid theory of gravitation that he had developed in 1912–13 in collaboration with the Zürich mathematician Marcel Grossman. This theory correctly demanded that all equations of physics be covariant under general continuous transformations of the space–time coordinates. This covariance was actually implemented for the geodesic motion of a point particle and the Maxwell equations. However, until the early fall of 1915, Einstein erroneously believed that he had proved that the field equations of gravitation could not satisfy the demand of general covariance.

In 1914, Einstein was nominated by Naunyn for his work on special relativity, diffusion, and gravitation and by Chwolson for his contributions to several domains of theoretical physics. The Report for 1914 noted vaguely that it might be a long time before the last word had been said about Einstein's theory of special relativity and his other work. For the 1916 award, the next one for which he was nominated, there was only one letter, from Ehrenhaft of Vienna, who proposed him for his work on Brownian motion and special and general relativity. The Report for that year noted, however, that the work on relativity was not yet complete.

The upswing started, slowly, with the nominations for 1917. De Haas proposed Einstein for the new theory of gravitation, quoting Einstein's explanation of the precession of the perihelion of Mercury. Warburg nominated him for the work in quantum theory, relativity, and gravitation. The third and last letter for that year, written by Weiss of Zürich, was the finest nomination for Einstein ever written. For the first time, we find an appreciation of the whole Einstein, whose work represents '*un effort vers la conquête de l'inconnu*'.[21] The letter first describes Einstein's work in stati stical mechanics centering on Boltzmann's principle, then the two axioms of special relativity, next the light-quantum postulate and the photoelectric effect, then the work on specific heats. It concludes by noting Einstein's experimental efforts: in 1915, Einstein and De Haas had performed experiments on the gyromagnetic ratio of the electron. The Report for 1917 referred to 'the famous theoretical physicist Einstein' and spoke highly of his work, but it ended with a new experimental snag – namely, that the red shift of spectral lines predicted by the theory of general relativity did not show up in the measurements of C. E. St. John at Mount Wilson. It concluded, therefore, that Einstein's relativity theory, whatever its merits in other respects, did not deserve a Nobel Prize.

For 1918, Warburg and Ehrenhaft repeated their earlier nominations; Wien and Laue independently proposed that the Prize be shared by Lorentz and Einstein for special relativity; Edgar Meyer of Zürich cited Einstein's work on Brownian motion, specific heats, and gravitation; and Stefan Meyer of Vienna cited his work on relativity (from now on, relativity refers to the special and general theories of relativity). The Report was almost identical with that of the year before.

For the 1919 award, Warburg, Laue, and Edgar Meyer repeated their earlier nominations. Planck nominated Einstein for general relativity, on the grounds that he had taken the first step beyond Newton,[22] and the Swedish physical chemist Arrhenius nominated him for Brownian motion. Perrin, Svedberg, and Gouy, all major contributors to experimental work on Brownian motion, were also nominated. The Report went into statistical problems in detail and included a discussion of Einstein's 1905 Ph.D. thesis, which dealt with the sizes of molecules, as well as his work of 1911 on critical opalescence. It noted, however, that Einstein's statistical papers were not of as high a caliber as his work on relativity and quantum physics and concluded by pointing out that 'it would appear peculiar to the learned world if Einstein were to receive the Prize for statistical physics . . . and not for his other major papers.' It was thought to be best to wait for clarification of the red-shift problem and for the solar eclipse of May 29!

For the 1920 Prize, Warburg repeated his earlier nomination; Waldeyer-Hartz, an anatomist from Berlin, and Ornstein of Utrecht cited general relativity; and a letter dated 24 January 1920, signed by Lorentz, Julius, Zeeman, and Onnes, stressed the theory of gravitation. The letter, which said that Einstein was in the first rank of physicists of all time, emphasized the successes of the perihelion motion and the bending of light and suggested that the red-shift experiments were so delicate that no firm conclusions could be drawn yet.

Niels Bohr added his voice too, citing Brownian motion, the photoelectric effect, the theory of specific heats, and, 'first and foremost, relativity

. . . advances of decisive significance for the development of research in physics.'[23]

Appended to the Report for 1920 was a statement by Arrhenius, prepared at the request of the Committee, on the consequences of general relativity. Arrhenius noted that the red-shift experiments still disagreed with the theory, and that criticism, much of it sensible, had been leveled from various sides against the bending-of-light results of the 1919 eclipse expeditions. The Committee concluded, therefore, that for the time being, relativity could not be the basis for the award.

In a brief, forceful note, Planck repeated his nomination of Einstein in 1921. De Haas and Warburg were also back. General relativity was cited in letters from Dällenbach (Baden), Jaffe and Marx (Leipzig), Nordström (Helsingfors), Walcott (Washington), and Wiener (Leipzig), Hadamard (Paris) proposed that either Einstein or Perrin be awarded the Prize, Einstein for his work on Brownian motion, quantum theory, and relativity, Perrin for his experimental investigations of Brownian motion. Lyman (Harvard) cited Einstein's contributions to mathematical physics.

Oseen, a theoretical physicist at the University of Uppsala, proposed Einstein for the photoelectric effect.

Eddington, who was among those who singled out general relativity, wrote that 'Einstein stands above his contemporaries even as Newton did.'[24] At this point, the Committee requested that one of its members, Allvar Gullstrand, prepare an account of the theory of relativity and that Arrhenius, another member of the Committee, do the same for the photoelectric effect.

Gullstrand, professor of ophthalmology at the University of Uppsala since 1894, was a scientist of very high distinction. He had graduated from medical school in 1890 and had become the world's leading figure in the study of the eye as an optical instrument. In 1960, the following was written of him: 'Ophthalmologists consider him to be the man who, next to Helmholtz, contributed more than anyone else to a mathematical understanding of the human eye as an optical system . . . While making these investigations, he discovered a number of widespread misconceptions about optical image formation, and, being a fighter, he devoted many of his later papers to an attempt to destroy these misconceptions'.[25] In 1910 and again in 1911, he had been nominated for the Nobel Prize in physics. In 1911, the initial suggestion of the Committee for Physics of which Gullstrand was a member from 1911 to 1929 and chairman from 1923 to 1929, was that the Prize be given to him for his work in geometrical optics. However, it turned out that the Committee for Physiology and Medicine was thinking of giving Gullstrand its Prize for his work on the dioptrics of the eye. Gullstrand declined the Prize in physics, and the Committee wrote another report, now including Gullstrand among the signers, suggesting Wien for the Prize.[26] Gullstrand thus became the only person both to decline and to receive a Nobel Prize.

Gullstrand's report, highly critical of the theory of relativity, was not a good piece of work. The summary found in the Report for 1921 states that 'the effects [of special relativity] which are measurable by physical means are, however, so small that, in general, they lie below the limits of experimental error.' Also beside the mark is Gullstrand's comment about general relativity, 'that it remains unknown until further notice whether the Einstein theory can at all be brought into agreement with the perihelion experiment [*sic*] of Le Verrier.' Gullstrand had fallen into the trap – he was not the only one – of believing he had shown that the answer for the perihelion effect depends on the coordinate system chosen for the calculation. He also expressed the opinion (more reasonable, though not very weighty) that it was necessary to reevaluate other, long-known deviations from the pure two-body Newtonian law using general relativistic methods before attempting to identify the residual effect. On 25 May 1921, he had presented a paper[27] on these considerations, a reprint of which was appended to his report. It was the first time he had ever published on relativity. It is no more than my guess that he may have become intrigued with general relativity because of one feature to which he had contributed a lot, albeit in a quite different context, namely, the bending of light.

The main points of Arrhenius's report on the photoelectric effect were first, that a Prize for quantum theory had just been given, to Planck in 1918, and second, that if work on the photoelectric effect were to be honored, it would be preferable to give an award to experimentalists. In the end, no Prize for physics was given that year.

The list of nominations for the 1922 Prize was even longer. Ehrenhaft, Hadamard, Laue, E. Meyer, S. Meyer, Naunyn, Nordström, and Warburg again recommended Einstein. There was a beautiful letter from Sommerfeld. There were also letters from de Donder (Brussels), Emden and Wagner (Munich), Langevin (Paris), and Poulton (Oxford). Brillouin wrote: 'Imagine for a moment what the general opinion will be fifty years from now if Einstein's name does not appear on the list of Nobel laureates.'[28]

Planck proposed that the prizes for 1921 and 1922 be given to Einstein and Bohr, respectively. Oseen repeated his nomination for the photoelectric effect.

The Committee asked Gullstrand for a further report on relativity and Oseen for a report on the photoelectric effect. Gullstrand stuck to his guns. He published a rebuttal[29] to the criticism leveled at his previous report by Erich Kretschmann, Privatdozent at Königsberg,[30] and appended a reprint of it to his new statement.

Oseen produced excellent analyses of Einstein's 1905 paper on the light-quantum and his 1909 work on energy fluctuations in black-body radiation.

The Committee proposed Einstein for the 1921 Prize. The Academy voted accordingly.

That is how Einstein got the Nobel Prize 'for his services to theoretical physics, and especially for his discovery of the law of the photoelectric effect.' That is also why Aurivillius wrote on 10 November 1922 that Einstein's award was not based on relativity.[1]

In his presentation speech on 10 December 1922, Arrhenius said: 'Most discussion [of Einstein's work] centers on his theory of relativity. This pertains to epistemology and has therefore been the subject of lively debate in philosophical circles. It will be no secret that the famous philosopher Bergson in Paris has challenged this theory, while other philosophers have acclaimed it wholeheartedly'.[31]

Bergson's collected works appeared in 1970.[32] The editors did not include his book *Durée et simultanéité: A propos de la théorie d'Einstein*. Einstein came to know, like, and respect Bergson. Of Bergson's philosophy he used to say, 'God forgive him.'

A further exchange between Gullstrand and Kretschmann settled their differences to their mutual satisfaction.[33]

In March 1923, Arrhenius wrote to Einstein[34] suggesting that he should not wait until December for his planned visit to Sweden but come in July, when he could attend a meeting of the Scandinavian Society of Science in Göteborg, on the occasion of the three-hundredth anniversary of the founding of that city. Arrhenius left to Einstein the choice of a topic for a general lecture but said that 'it is certain that one would be most grateful for a lecture about your relativity theory.' Einstein replied that he was agreeable to this suggestion, although he would have preferred to speak on unified field theory.[35] On a very hot day in July, Einstein, dressed in a black redingote, addressed an audience of about two thousand in the Jubilee Hall in Göteborg on basic ideas and problems of the theory of relativity.[36] King Gustav V attended the lecture and had a pleasant chat with Einstein afterward.[37]

The second coming of Planck's constant

Why did Einstein not get the Nobel Prize for relativity? Largely, I believe, because the Academy was under so much pressure to honor him. That so many letters were sent on his behalf was never the result of any campaign but rather of leading scientists recognizing the value of his work. It is understandable that the Academy was in no hurry to award the Prize for relativity until the experimental issues had been clarified. It was also the

Academy's bad fortune not to have among its members in those early years anyone who could competently evaluate the content of relativity theory. Oseen's proposal to give Einstein the award for the photoelectric effect must thus have come as a relief of conflicting pressures.

Was the photoelectric effect worth a Nobel Prize? Without a doubt.

In 1902, Lenard had studied this effect using a carbon arc light as a source.[38] Varying the intensity of his light source by a factor of the order of 1000, he had made the crucial discovery that the energy of the electrons liberated from a metal surface irradiated with his arc lamp showed 'not the slightest dependence on the light intensity'. As for the variation of the photoelectron energy with the light frequency, one increases with the other. Nothing more was known in 1905.

That was the year in which Einstein proposed the following 'simplest picture' for the photoelectric effect.[39] A quantum of light gives all its energy to a single electron. The energy transfer by one such quantum is independent of the presence of other light-quanta. An electron ejected from the interior of an irradiated body will, in general, suffer an energy loss before it reaches the surface. If E_{max} is the electron energy for the extreme case in which the energy loss is zero, then the relation proposed by Einstein, in modern notation, is as follows:

$$E_{\mathrm{max}} = h\nu - P \qquad (1)$$

where ν is the frequency of the incident monochromatic radiation, and P is the work function – that is, the energy the electron needs to escape from the surface. This relation is independent of the light intensity and therefore explains Lenard's result. But it does much more.

Eqn (1) represents the second coming of Planck's constant, h.

Before Einstein wrote down this equation, there was only one other experimentally verifiable relation involving h, namely, Planck's equation of 1900 for the energy density per unit volume, $\rho(\nu, T)$, of black-body radiation at frequency ν and temperature T:

$$\rho(\nu, T) = \frac{8\pi h\nu^3}{c^3}\,(e^{h\nu/kT} - 1)^{-1}, \qquad (2)$$

where k is the Boltzmann constant and c is the velocity of light. Planck's bizarre justification of Eqn (2) need not concern us here. Suffice it to note that, in deriving his formula, he was unaware that light comes in quanta, or photons. In fact, Planck, together with many other leading physicists, including Bohr until 1925, remained opposed to the idea of photons long after 1905 when Einstein first introduced it.

When Einstein wrote down Eqn (1) in 1905, he made three predictions that were not verified until a decade later: that E_{max} increases linearly with

ν; that the slope of the energy–frequency curve is a universal constant, independent of the nature of the irradiated material; and that the value of this slope is Planck's constant, as determined from the radiation law.

It is thus evident that the photoelectric effect did deserve a Nobel Prize. In fact, the order of the awards made for quantum physics was perfect: first Planck, then Einstein, then Bohr. It is a touching twist of history that the Committee, conservative by inclination, honored Einstein for the only paper which he himself described – with good reason, I believe – as revolutionary.

References

SAS: Royal Swedish Academy of Sciences

1. C. Aurivillius, letter to A. Einstein, 10 November 1922.
2. *Nobel Lectures in Physics, 1922–41*, Elsevier, New York, 1965.
3. Report on the Joint Meeting of the Royal Society and the Royal Astronomical Society, London, 6 November 1919. *Observatory*, **42**, 389–98.
4. A.S. Eddington, *Observatory*, **42**, 119–22, 1919.
5. I. Newton, *Opticks*, 1704.
6. A. Einstein, *Annalen der Physik*, **35**, 898–908, 1911.
7. A. Einstein, *Sitzungsberichte der Preuss. Ak. der Wiss.*, **1915**, pp. 831–39.
8. H.A. Lorentz, letter to P. Ehrenfest, 22 September 1919.
9. M. von Laue, letter to A. Einstein, 18 September 1922.
10. R. Nadolny, in *Les Prix Nobel*, pp. 101–2, Imprimerie Royale, Stockholm, 1923.
11. C. Stumpf, letter to A. Einstein, 22 October 1919.
12. C. Kirsten and H.J. Treder, *Albert Einstein in Berlin*, Vol. 1, Akademie Verlag, Berlin, 1979.
13. I. Einstein, letter to Sederholm, 6 April 1923.
14. B. Nagel, The discussion of the Nobel Prize for Max Planck. In *Science, Technology, and Society in the Time of Alfred Nobel. Nobel Symposium, Karlsoga 1981*. Almquist and Wiksell, Stockholm, 1982.
15. A. Einstein, letter to SAS, undated, 1918.
16. W. Ostwald, letter to SAS, 2 October 1909.
17. W. Ostwald, letter to SAS, 21 December 1911.
18. W. Ostwald, letter to SAS, 30 December 1912.
19. E. Pringsheim, letter to SAS, 12 January 1912.
20. W. Wien, letter to SAS, January 1912.
21. P. Weiss, letter to SAS, 21 January 1917.
22. M. Planck, letter to SAS, 19 January 1919.

23. N. Bohr, letter to SAS, 10 November 1920.

24. A.S. Eddington, letter to SAS, 1 January 1921.

25. M. Herzberger, *Optica Acta* **7**, 237–41, 1960.

26. B. Nagel, pers. comm. 1981.

27. A. Gullstrand, *Ark. for Mat. Astr. och Fys.* **16** (8), 1–15, 1921.

28. M. Brillouin, letter to SAS, 12 November 1921.

29. A. Gullstrand, *Ark. for Mat. Astr. och Fys.*, **17** (3), 1–5, 1922.

30. E. Kretschmann, *Ark. for Mat. Astr. och Fys.*, **17** (2), 1–4, 1922.

31. *Nobel Lectures in Physics, 1901–21*, p. 479. Elsevier, New York, 1967.

32. H.L. Bergson, *Oeuvres*, ed. A. Robinet, Presses Universitaires de France, Paris, 1970.

33. E. Kretschmann, *Ark. for Mat. Astr. och Fys*, **17** (25), 1–4, with an added comment by A. Gullstrand, 1923.

34. S. Arrhenius, letter to A. Einstein, 17 March 1923.

35. A. Einstein, letter to S. Arrhenius, 25 March 1923.

36. A. Einstein, *Grundgedanken und Probleme der Relativitätstheorie*. Imprimerie Royale, Stockholm, 1923.

37. J.A. Hedvall, letter to H. Dukas, 19 November 1971.

38. P. Lenard, *Annalen der Physik*, **8**, 149–98, 1902.

39. A. Einstein, *Annalen der Physik*, **17**, 132–48, 1905.

7

*Helen Dukas, in memoriam**

Tov shem meshemen tov, wejom hamawet mejom hewaledot. A good name is worth more than precious oil, and the day of death more important than the day of birth.

We called her Helen. Einstein called her Dukas, and referred to her as 'die Dukas'. Her first name was Helena. She was born in Freiburg im Breisgau, in Germany, on 17 October 1896. Her father, Leopold Dukas, a native of Sulzburg, a small town in the Black Forest, was a wine merchant who also produced the famed Schwarzwalder Kirschwasser in a government-controlled distillery behind the house in Freiburg. Her mother, Hannchen Dukas-Liebmann, hailed from Hechingen in Hohenzollern, the same town where Elsa Einstein, Albert Einstein's second wife, was born. From the Hechingen days Elsa knew Hannchen and also Helen's grandmother, who was known in the community as 'Tante Lisette'. These acquaintanceships of Elsa were to seal Helen's destiny.

The Dukases had seven children. The first two were girls; then there were two boys; then, after quite a few years, three more girls. Helen was the eldest of these three. She used to refer to herself as 'the oldest of the youngest'. From the age of six to fifteen she attended the *Höhere Töchterschule* in Freiburg. That was all the formal education she received. In 1909 her mother died of tuberculosis, '*erschöpft von den sieben Kindern*', exhausted from the seven children. Celine, Helen's oldest sister, took charge of the running of the household until she married. Thereafter it was Helen's turn; she had to leave the school on the Holzmarktplatz without having completed the curriculum.

Helen stayed in Freiburg until 1921, taking care of her brothers and

* Remarks at the memorial ceremony, Institute for Advanced Study, 15 March 1982.

sisters (her father died in 1919), and also starting her first job, as kindergarten teacher. Then she went for a year to Munich, as governess in the home of Raphael Straus, an uncle of one of Einstein's assistants in the 1940s, Ernst Gabor Straus. (When, in 1944, Ernst introduced himself to Helen, she told him at once that she had been at his *brith*, the festive day of circumcision.) Thereafter she moved to Berlin, where some of her family had settled meanwhile. For some years she worked as secretary for a small publishing house, until that company went out of business and she was without a job.

It was during that period, on 11 April 1928, to be precise, that Elsa Einstein asked Helen's older sister Rosa if she knew someone suitable to be secretary to her husband. The two women knew each other through '*Jüdische Waisenhilfe*', an organization for aid to Jewish orphans. Rosa was its executive secretary, Elsa its honorary president. The two women had become friendly especially because of the Hechingen connection. In response to Elsa's question, Rosa proposed Helen. Elsa thought this was a fine idea. She had seen Helen a few times, when the latter came to the Einstein apartment on the Haberlandstrasse to bring letters which she, Elsa, had to sign. She asked Rosa to have Helen visit her. That same day Rosa called Helen and told her the news. Helen's reaction was: '*Du bist verrückt geworden, so was kann ich nie tun*', 'You have gone mad, I can never do something like that.' Nevertheless, the next day Helen visited Elsa, who plied her with tea and cookies and urged her at least to try.

The day after Helen came back. It was Friday the thirteenth. Einstein was bedridden at that time, recuperating from heart disease. Elsa brought Helen to the bedroom. Einstein greeted her with a friendly smile and an outstretched hand, saying: '*Hier liegt eine alte Kindsleich*', here lies an old child cadaver. He immediately asked her to call the *Kultusministerium* in connection with his duties as member of a committee of the League of Nations. Helen, brought up as '*authoritätsglaübiges deutsches Schulmädchen*' (German schoolgirl with religious belief in authority), did so with trepidation. All went smoothly, however, and 'for the first time I experienced what magic effect the name "Einstein" had.' Next she typed some letters. Then Einstein told her he thought they would get along very well. 'Then the remaining feelings of inferiority left me, even though in the next twenty-seven years I never lost one iota of respect nor a certain shyness.'

In the course of time Einstein developed a growing appreciation and respect for Helen's capabilities, together with a strong affection that never went beyond the fatherly.

I turn to the next great event in Helen's life: her three months' trip to California, together with the Einsteins and Walther Mayer, Einstein's assistant (December 1930–March 1931). Helen kept a diary which she gave me to read. Its bulk is devoted to the California trip. Otherwise it is, alas, quite sketchy. She tells of the hotel in Antwerp where she had her first room with bath ever; of playing ping-pong with Einstein on board the *Belgenland*; of Jimmy Walker presenting the key to New York City to Einstein; of an old woman who broke through a police cordon, pressed Einstein's hands, and said, 'Now I can die in peace'; of Helen Keller first addressing Einstein then touching his skull and his face, with Einstein in tears; of the relief of having to mend socks in a New York hotel; of a visit to Cuba; of attending the Tournament of the Roses in Pasadena; of the historic first meeting between Einstein and Michelson; of meeting Charlie Chaplin at a dinner in his house attended also by the Einsteins, Marion Davies, and William Randolph Hearst; and of being taken to the world premiere of *City Lights* at the Los Angeles Theatre. All this, written in fine style and with great wit, represents an important eyewitness account of uncommon events that occurred wherever Einstein went.

In December 1932 the Einsteins left for what was planned as another trip to California, without Helen this time. Hitler came to power the next month, and they never returned to Germany, but they did go back for a last European encampment, in Belgium. Helen joined them there, and in October 1933 she, the Einsteins, and Mayer sailed for America. They were taken off the ship as it entered New York harbor and brought by launch to the Battery in downtown Manhattan. There Edgar Bamberger and Herbert Maass, trustees of the Institute for Advanced Study, were waiting and took them by car to the Peacock Inn in Princeton. A few days later the Einsteins and Helen moved into a house on 2 Library Place. In the autumn of 1935 they moved to 112 Mercer Street. Soon thereafter Elsa's daughter Margot joined them there.

After the death of Elsa in 1936, Helen took full responsibility for managing the household, together with her continuing secretarial duties. Einstein's sister Maja, who had joined the household in 1939, was bedridden from 1945 until her death – in the Mercer Street home – in 1951. It was Helen who arranged for the medical attention.

We have Helen's own account of her daily activities in Princeton from a letter she wrote to Carl Seelig in the early 1950s. After having described a typical day in the life of the Professor she turns to '*meine Wenigkeit*,' my humble self:

> [To describe my own days] is much more difficult, but then that is less important. It is more complicated because I have so many duties, I mean all kinds of things. I also try, as much as possible, to stick to an organized routine, but that is not always easy. At about 8.30 I come downstairs, a bit later on the four days on which I have household help, something which is not so common in this country. Then I make break-

> fast or at least coffee and tea etc. On the days on which our Irish help is here I go to town after breakfast, first to buy food, then to go to the bank and to the post-office or to do what else is necessary. Around eleven, sometimes later, I get back, feed the dog, open the mail, throw – *Gottlob* – most of it in the waste-paper basket, go to the kitchen and prepare lunch. In between I quickly write letters if that is necessary, else I do something about the house. After lunch comes the sacred rest – but on three days per week the dishes get done first. From four o'clock on there is either dictation of letters or the writing of letters of the previous day or replies which I attend to myself. Sometimes there are visitors for tea. Six o'clock: back to the kitchen for the evening meal. In the evenings there is either dictation, or the typing of letters, or reading, or a movie if there is something of interest. Of course there are also visits from friends or to friends. Now and then there is a trip to New York, but not often, because it is strenuous and because I first must carefully prepare my departure. Usually something comes in between at the last moment. I have of course given only a crude outline. On top of all this comes my special bane, the telephone, the doorbell, the curious, the reporters, the crazies, etc. Telegrams and express letters have long since failed to make any impression. What I hate most is the filing of letters, especially because I have so little space. I have filing cabinets even in the hallways and there are books everywhere, innumerable crates in the basement. I have often wished that Gutenberg had never lived! I do not know if you can make yourself a picture. I appear to myself as the farmer's wife who is asked by someone making a statistics how she and people like her spend the day. She enumerates her duties: to cook, to wash, to iron, to milk the cows, to feed the chicken, to clean, to sew, to work in the garden etc. Finally the man asks: 'What do you do with your free time?' To which she replies: 'Then I go to the John', '*dahn gehe ich aufs "Oertli"*.' With me it is not quite as bad, but still the comparison does hold a little bit.

Two stories I owe to Helen may serve to illustrate the bizarre happenings that would occasionally take place on Mercer Street. One day, at about ten in the morning, the bell rang. As Helen opened the door she saw a man who said to her: 'I am Judge T—— and would like to see Professor Einstein.' Helen called Einstein, who came to the door. When the judge saw him he passed out. After he came to he had to be helped into a taxi. The next morning the judge returned, quite contrite. He explained that he had been quite intoxicated on his previous visit, in spite of the early hour, the reason being that he was in town for the University Reunion.

The second story concerns an event on a day in 1939. The bell rang. Helen went to the door. She saw a well-dressed man who said 'Madam, I have come to the United States to see the World Fair, Professor Einstein, and the Grand Canyon.'

For those left behind there are no better ways of coping with death than to have loved well and to have served well. Helen was no exception. Shortly after April 1955 she threw herself with renewed vigor into a major task, the creation of an archive for Einstein's papers and letters. At the time of Helen's death these archives consisted for roughly 5000 pages of published and unpublished manuscripts, 3000 pages of notebooks and travel diaries, and 52,000 pages of correspondence, of which 29,000 are of a non-scientific nature. The correspondence has been abstracted and catalogued both alphabetically and chronologically. Eminent men, especially Gerald Holton and Martin Klein, helped in scientific matters of which she would smilingly confess her complete incompetence. Otherwise nearly all the labor was hers. My (earlier) book on Einstein bears ample testimony to her achievements and will help, I hope, to secure her a well-deserved place in the history of physics. Here I shall only add two comments.

Helen's most important single contribution to the archives, in my judgement, was the result of a correspondence between her and Frau Janke, David Hilbert's 'Dukas'. After Hilbert's death in 1943 the Hilbert–Einstein correspondence had been given to the University Library in Göttingen, where it lay forgotten. Because of an initiative by Helen, the letters were recovered by Frau Janke and sent to Princeton. This correspondence contains a crucial exchange in November 1915 without the knowledge of which the history of general relativity cannot properly be written. Helen's talents for knowing where to look and whom to ask have enriched the archives very substantially.

My second remark may serve as but one illustration of Helen's astounding familiarity with the contents of the archives. At one point I was interested in the contacts betwen Einstein and Poincaré. So I went to Helen and asked her if there had ever been an exchange of letters between them. 'No,' she said. That simple answer saved me considerable labor. Then she said that there was of course this exchange about Poincaré with that Swedish mathematician. 'Which mathematician?' I asked. 'Mittag-Leffler,' she replied. 'Can I see those letters?' I asked. 'Of course,' she said. I read the letters and caught a new glimpse of Einstein's feelings about Poincaré. As I returned the letters to her, she commented that there was of course that lecture by Poincaré in Berlin in 1910. 'How do you know about that?' I asked. 'Because it is in the first chapter of Moszkowski's Einstein biography,' she said. I looked up that book. There it was, just as she told me.

I believe that death was on Helen's mind in the last year of her life, during which she said to me several times that only helping me with my book kept her alive. She certainly did prepare for death. Article (13) of Einstein's last will contains the stipulation: 'To deliver and turn over to

Hebrew University any funds or specific property held in trust, at any time, upon the written direction of the said Helen Dukas during her life-time . . .' From 1 January 1982, the Estate of Albert Einstein had transferred all papers, publications, manuscripts, and all other property to the Hebrew University.

Helen's was a rich life. She did her work with competence and good cheer. She read voraciously. She was intensely Jewish, though not religious. Though not free of what I call Fallen Hero Syndrome, which causes confusion between greatness and immaculacy, she was fundamentally capable of that touch realism that so often marks women of quality and substance. She was blessed with the abilities to give and to receive affection. She was lucid until the end.

In the summer of 1981 I wrote the following lines in the preface to my biography of Einstein: 'No one helped me more than Helen Dukas, more familiar than anyone else at this time with Einstein's life, trusted guide through the Einstein Archives in Princeton. Dear Helen, thank you, it was wonderful.' A few weeks ago I added the following sentence in proof: 'I have left the text of this Preface as it was written before the death of Helen Dukas on February 10, 1982.'

Note. The details of how Helen started to work for Einstein are described in her letter to Carl Seelig, now in the Wissenschaftshistorisce Sammlung, ETH Bibliothek, Zürich, Hs. 304:118. Her letter to Seelig describing a day in the life on Mercer Street is in that same collection, Hs. 304:133.

8

Samples from Die Komische Mappe

Komisch, adj. Strange, funny, queer, odd; pathetic. *Langenscheidt's Dictionary*

1. Preamble

Men and women who are in the public eye have to pay the price of receiving numerous messages from strangers. Their contents vary, from congratulatory notes to requests for help or additional information to thoughtful commentaries to lunatic utterances to expressions of hate, or threats. Those in receipt will answer some, throw others directly away, or keep some in files which may carry such picturesque names as the 'crackpot file'.

It is a safe bet that among scientists no one received more such letters than Einstein. I have no idea how many of them were sent. In any event, over 600 have survived and are now in the Einstein Archive.* Einstein himself referred to this collection as *die komische Mappe*. Only very rarely does one find a copy of a reply.

As is apparent from the quotation above, the German *komisch* has a variety of inequivalent English translations. I shall not single out any one of these as my translation of choice. As the reader peruses excerpts from these letters, which follow, he or she will observe that now one, then another, of these adjectives will best apply. It will further be noted that there are some instances where 'none of the above' conveys the spirit of some or other letter.†

It would make little sense to reproduce all these documents, if only

* Permission to quote from these has been granted by the Albert Einstein Archives, the Hebrew University of Jerusalem.

† I have also seen the translation 'the curiosity file', which has something to commend itself, taking the key word in the Dickensian sense. *Note*: In the *Oxford English Dictionary* one finds sixteen inequivalent definitions for 'curiosity'!

because many of them are of little concern. It seemed to me that it may be of some service to the general public, however, to give here parts of some of these epistles that appeared to me to represent best the variety of categories mentioned at the beginning. I have refrained from editorializing since it seems to me that if these letters are of interest at all they should speak for themselves. As a result, my presentation will be seen to have a hop-skip-jump quality – which in fact accurately reflects my impressions as I waded through this material myself. For this reason I have refrained from categorizing these documents, and rather present them in chronological order of receipt. I hope that the resulting essay – which I do not consider to be a literary masterpiece – may serve to illuminate a less familiar facet of the life of a most unusual man.

I have thought it best to refrain from citing authors, since these are by and large good citizens (only a few are not so good), not known to anyone but their near kin. For those interested in my sources, all those are in the Einstein Archive, as previously mentioned. Following that institution's cataloguing system, I shall refer to the material used by reel number, followed by folder number and document number. Thus (31,9) 816 means: reel 31, folder no. 9, document no. 816.

2. Envelopes

In some cases only envelopes have been preserved.[1] Samples:

From Brooklyn: 'To the beloved Albert Einstein, professor and messenger of God and servant of humanity'.

From Manila: 'To Doctor Albert Einstein, Institute of Living, Princeton'.

From Troy, New York: 'To Doctor Albert Einstein, Chief engineer of the universe'.

From Sweden: '*Herr Oberwirrkopf*, Mister Albert Einstein, Esq.' ('Mr. Superscatterbrain . . .')

From Durban, South Africa: 'Professor Albert Einstein, Master tailor of clothes'.

Provenance unmarked: 'Mister Albert Einstein, scientist, some place in the United States, I think some college in the United States.' The envelope shows a postmaster's mark: 'Not at University of Chicago, try Harvard.' Evidently it reached its destination eventually.

3. Letters from the Berlin period

Einstein held academic positions in Berlin from 1914 until 1933. His world fame began in 1919 when European as well as American newspapers started giving extensive coverage to the confirmation of his prediction – a consequence of his general relativity theory – of the amount starlight is bent when passing the Sun. There can be no doubt that letters from far and wide must have begun to reach him shortly thereafter. The first letters found in the *komische Mappe* date from only a decade later, however. It is probable that here one sees the hand of Helen Dukas, who was Einstein's secretary from 1928 until his death, and who was instrumental in getting the *komische Mappe* together in the years thereafter.

I became aware of this collection only after Helen's death in 1982. Thus I was never in a position to discuss these letters either with her or with Einstein himself. Now to the letters.

An early one[2] begins with a supplication for 'the favor to kneel with tears in the eyes before the new prophet on the throne of science'. It is a request for assistance from a Rumanian Jew to enable him to study at a German university. Einstein must have replied (that letter is lost), for some time later the same person wrote[3] that he is 'proud to have a letter from the great son of his people, that the great Einstein has taken the trouble for a little *Ostjude*'. I do not know what became of the poor fellow.

Next, a person from Clifton, New Jersey, asks Einstein for information[4] about the love life of his (E.'s) parents 'for a study aiming to show that, when man and wife sleep in separate beds and have less marital contact, they have better prospects for having children with better mental and physical abilities.' A clergyman from New York informs Einstein[5] that he has composed an 'Einstein March'. A man from Jerusalem asks[6] for his opinion as to whether it could be possible that the walls of Jericho had fallen (in the days of Joshua) as a result of synchronized movements resulting from the steady, rhythmical marching of the Jewish people, documenting his assertion with numerous biblical quotations.

A quite unusual letter[7] was sent by a Rumanian architect:

> I am only a little technician . . . I mean well regarding you, but only wish to liberate you from an error. As a result of ovations by the masses you seem to overestimate the value of your personality. That is a tremendous error! You should know that in our days our world, especially the Americans, only needs propaganda and advertising in order to elevate someone from one day to the next. You, as a clever man, should know that, since you yourself are a propagandist and advertising personality . . . Why don't you rather retire in silence and fertilize your mind with a new idea . . . He who has reached heights must suffer his decline . . . You ought to know that the whole world is not so stupid as you may imagine from the homage you have received . . . there are still some, perhaps not many, who cannot be blinded or seduced by the noises of propaganda and sensation . . .

This is one of the rare letters to which Einstein replied: 'Thank you for your fatherly admonition. Your letter gets a place of honor next to one by a flatterer.'

A German living in Brazil writes thoughtfully[8] that 'there are still other kinds of Germans than those who caused you such sad experiences'. From New York comes the question: 'Would it be reasonable to assume that it is while a person is standing on his head – or rather upside down – that he falls in love or does other foolish things?'[9] In a corner of this letter, in Einstein's handwriting: 'To fall in love is by no means the most stupid thing man does – gravitation cannot be held responsible, however.'

This concludes my selection from the 130 letters in the *komische* folder which Einstein received in his Berlin days. On 17 October 1933, Einstein, accompanied by his wife, Helen Dukas, and a scientific assistant, arrived in the United States. That same day they went on to Princeton, which was to be Einstein's home for the rest of his life.

Most of the letters in the file date from his American years.

4. Letters from the Princeton period

1935. A man from New York sends his visiting card 'which I am selling to you', and signs his name, followed by 'Esoteric Mystificationist, Scientist Extraordinary'.[10] From the monthly *Listening in*: 'We are asking ten outstanding Americans: What believe you of the Devil?'[11] A Jew from Košice, Czechoslovakia, asks Einstein's help[12] in obtaining immigration certificates for Palestine for two daughters 'who unfortunately I cannot marry off here'. There are two sad letters from mentally ill people, one in the Norfolk State Hospital, Nebraska.[13] (More such came in later years. I shall not quote from any of these.)

1936. From a lady in New York: 'Isn't the fourth dimension a matter of diffusion and intensity, the distillation of things into their original essences? It has to return to the point. Isn't the fourth dimension – the point?'[14] From the Shoe Club, Inc., New York, an organization comprising tanners, manufacturers, and merchants: 'The Shoe Club feels that a collection of shoes that have been worn by men of renown would be an inspiration . . . Would you be kind enough therefore to send us one of your old shoes, preferably the right shoe? This shoe will be permanently preserved and on display at our headquarters at the Hotel McAlpin.'[15]

1937. On 20 December 1936, Elsa, Einstein's second wife, died in Princeton. Like everything concerning Einstein's life, this news was reported in the world press. As a result, in early 1937 Einstein received two most unusual letters containing marriage proposals. Both are well phrased; the handwritings are handsome. Their contents are touching.

The first[16] is from Astoria, Long Island:

> I read you are now a widower. I [am] a widow of the same faith as you . . . How much brighter my existence would be if I could persuade you that we both should be happier were we to unite in cheering each other's daily path . . . I for my part will do all in my power to render your life happy and prosperous, and to shield you, as far as may be, from trouble and anxiety, if you will only give me the right to do so.

The second one[17] (in German) is from Vienna, from a widow who knew Elsa personally:

> A secret voice in my innermost soul (which rarely deceives me) tells me that I must devote my life to you . . . Please do not believe that I, out of pure vanity, have the ambition to become *Frau* Einstein . . . It is rather my deepest wish to give professor Einstein the most beautiful evening of life . . . I am not a woman of the world with blood-red lacquered finger nails. Many people find me beautiful, interesting, and engaging. Most important is, however, that I have a pure soul, a sunny disposition, and a feeling heart . . . Perhaps I am acting against good taste [in writing to you].

In 1944 another such tasteful letter arrived.[18]

It pains me to find these letters filed under the heading *komisch*.

More from 1937. A man from Texas sends[19] a manuscript 'Sex in Celestial Objects'. Comets are seen as 'analogous to the male spermatozoa'. A planet may be drawn into 'the seething mass of the great mother [the sun], later causing a parturient eruption by which a new planet is born'. Jupiter may 'break away with her cometary attendants, to set up a solar household of her own'.

A touching letter[20] from a woman from the Philadelphia area:

> From the standpoint of the universe and the ages my life, I realize, is of slight importance. But from the standpoint of my husband our young son and others whom my life affects it is exceedingly important that I develop a strong faith in the worth of human effort . . . For my spiritual need, I have wished to catch a little fire from your great soul's torch . . . A few words from you out of your own feeling and experience . . . would be of tremendous help to me.

No evidence, alas, that such words were sent. Another example of a letter which hardly can be called *komisch*.

1938. A Western Union telegram from the firm Young and Rubicam: 'Will you please wire collect permission to mention your name on our Ben Bernie Show in the following connection – Lew Lehrer talking about his son says – He's a little Einstein – a mental phenomenon etc.'[21]

A letter from Budapest.[22] Someone requests a meeting with Einstein in order to prove, 'however phantastic that may seem,' that he has found the physical law used by Stradivarius to build his violins.

From 1938 dates the earliest hate letter[23] I have found in these files:

'We certainly could do beautifully without you. You are just as undesirable here as in Germany. It's you and your kind that will bring trouble here, to your race, in the not very far off future. I have many Jewish friends, but not your type, thank goodness. Most of them came here with nothing, and found freedom and happiness, and unlike you, are grateful for it. Don't spoil the country for them, they love it.

'An admirer of the Jew in America, the right kind.'

This letter is most probably a reaction to Einstein's signing of an appeal by a group of Princeton professors to lift the US weapons embargo on arms to the Spanish (anti-Franco) government, news widely reported in the daily press.[24]

A letter from the Bronx,[25] asking Einstein 'to tell me your secret of success so that I and many others may use it as a foundation for our betterment and attainment toward that which we seek.'

From North Carolina: 'I have evolved a simple mathematical formula which explains all phenomena . . . that what I have is not a theory, but rather it is The Law; The Fulfillment.'[26]

A telegram[27] from Western Union:

> Understand you have issued statement declaring astrology a superstition. Because of your rating as a scientist and because no legitimate scientist ever condemns anything of which he is in complete ignorance we assume you have completely investigated the subject. Will you please advise us the number of years your investigation required and the number of assistants necessary to reach your stated conclusion. Starr Publishing Co. 227 West 13 St., New York.

1939. There are remarkably few letters from that year. Here are three items only.

From Los Angeles:

> Since you have found out that the church stands for truth, I hope that you will make a full surrender of your life to Jesus Christ and unite with the church of your choosing. I am inclosing an application blank which we are using in the Immanuel Presbyterian Church of Los Angeles and ask that you change the name and write in the church of your choice. I trust that I may have the pleasure of hearing from you as to your decision.[28]

There are many more such letters from others, begging Einstein to become a Christian.

Signature unreadable, place of origin not recorded:

> Please be advised that I have had a solution for the refugee problem for over a year and find it difficult to forget the idea, and am writing to you in the hopes of either getting a good reason why it cannot work or can work. Perhaps your answer will be in between. I shall take the answer as final since it comes from one of the foremost minds of this age.
>
> In the first place I must ask you to be unbiased, unprejudiced, for the idea when first heard sounds very fantastic. However, to my limited knowledge, it seems to me that it can be worked out on a practical basis.
>
> The plan is to construct a gigantic 'raft' 10 miles square in the Atlantic Ocean on which will be built homes. An advantageous material to supply buoyancy is lumber. Trees would be squared and kiln-dried to eliminate free air space and reduce the weight of the timber. They would be interlocked, bound together by bolts and set horizontally in layers so that only a small amount of lumber would come in contact with the water. I believe this arrangement would render the 'raft' waterproof, and fire-proof since there can be no draft to aid combustion. It is also sink-proof and bombproof.
>
> Because of its large area I do not believe that it would 'rock.' Soil would be put on the surface to grow wheat, corn and other crops and also to support live-stock. Drinking water would be distilled from the ocean and the remaining minerals would be used to fertilize the soil. It would take a very large area to distill enough water for the populace so there would be condensation tanks on the roofs of all buildings. This would serve a two-fold purpose in that the rooms would be cooled due to evaporation of the water on the roofs by the sun.
>
> There would be no need for a heating system since the floating city can be towed south in winter and north in summer. This advantage would eliminate seasonal businesses and the necessity for heavy clothes.[29]

From Western Union again: 'Interested in reading that professor Einstein has taken up dancing as a relaxation. Appreciate corroboration by wire collect postal telegraph. Many thanks. Edwin Cox Publishers Syndicate. 527 Fifth Ave New York City.'[30]

1940. From London from an RAF pilot: 'I hope we will win this war so that you may go and see your native country again.'[31] Nothing *komisch* about that.

From New York: 'Your attire – or lack of it – should be regarded by the authorities of Princeton University* as an insult to that institution. It stamps you as unfit to associate with socially decent people.'[32]

On 1 October 1940 Einstein was sworn in as citizen of the United States. Several people sent him letters of congratulations. The one I like best says: 'My heart swells with elation at the acquisition of a new "brother" Yankee whom I can admire and respect.'[33]

1941. A request[34] for rights to a film about Einstein 'stripped of fanfare or

* There is still some misunderstanding here: the Institute for Advanced Study is not part of Princeton University.

even subtle propaganda . . . I have taken the liberty of "feeling out" one of the major Hollywood film companies about the possibility of such a picture. It was received with splendid approval.' Just the kind of thing Einstein did not like, even without fanfare; no reply extant.

1943. 'My brother who is now 16, refuses to get haircuts. He is an admirer of yours and replies to urging that maybe he will grow up to be an Einstein. I have wagered that as a youth you got haircuts in due season and I will appreciate it greatly if you can let us know whether you did or not.'[35]

1944. An aide to a Maharajah requests an autographed photograph for His Highness.[36]

1945. From New York, an offer to enter a home for the aged.[37] From Bombay: A man who has made the discovery that the sun is not hot requests that Einstein proposes him for the Nobel prize in physics.[38]

Meanwhile atomic bombs have been dropped on Hiroshima and Nagasaki.

A child notes: 'Dear Professor. I respect your opinion. You tell us what to do in the atomic age and how to do it.' Signed: '6 years young.'[39]

From New York: 'Look here Doctor Einstein! Genuine Christian Americans resent your arrogance and European ideas . . . It will be well to keep your mouth shut on subjects you obviously know so little about.'[40]

From Cleveland: 'You'n I know, Ernie, that the earth ain't no more a globe, it was flat . . . Making you the Big Wig, with more hair than brains, makes you feel as if you do rate . . . You sure know how to play to the grandstands and strut your stuff, you old stiff.'[41]

1946. From New York: 'Dear Sir, What your Jewish scientists are doing through the air to Catholic people, using all your methods and inventions, is the worst crime in all World History. When the Truth will come out, all Christians will know the right reason for Hitler's madness killing of Jews.'[42]

Also from New York. 'Your advocacy of an agreement with Russia on a "grand scale" is typical logic which might be expected from a dumb Jew bastard like you . . . All the world wants from you is nothing, and plenty of it. Back into your hole, bum.'[43]

From Akron, Ohio: 'Why hunt for a world government? Our Creator gave us that; and all His natural laws that we must obey or perish.'[44]

From Germany, concerning Einstein's proposal to create a world government: 'That government should consist of psychiatrists.'[45]

From somewhere in the United States: 'You are the prince of idiocy, the count of imbecility, the duke of cretinism, the baron of morons and the king of stupidity and of morons.' Signed: 'A scientist.'[46]

Also from the USA: 'Erotic pleasure plays an important part – the most important role in life – [for] the reproduction of the species. Why shouldn't it play an important part in the relations of men with each other? Please give this the benefit of your wisdom.'[47]

From Chicago: 'You should resign your professorship as penance for the loss of lives of millions of innocent Jews you have contributed in murdering, and by taking the street sweeping job in Princeton.'[48]

1947. From East York, Ontario: 'Most honourable and distinguished Son of God . . . I would be sincerely honoured to wash your feet.'[49]

A postcard from Boston: 'You will immediately *stop* calling *space* curved.'[50]

From Romania: 'It would make me most happy if you could send me six pairs of nylon stockings.'[51]

1940. From New York: 'I don't have a baby. My wish is to adopt a baby, and I don't have money. Big Professor, please help me.'[52]

From Philadelphia: 'I must speak to you alone. I am the successor to Jesus Christ. Please hurry.'[53]

From Tokyo: 'Please give me your Golden Saying or guide of studying science for the egg of scientist. I hope that you are received my best respect.'[54]

1949. From Palo Alto, California: 'I am the prophetess of God for this era . . . But I have no money left now. I cannot get anyone to listen to me. I have ideas that need great men's minds to understand them.'[55]

From Hamburg: 'You are a great scientist, but already an old man. Would you like to be young again, Herr Professor? I can propose a method which will make all people young again, without operation . . . I should like to request to treat this matter confidentially, I do not want to make all of humanity young again, only a few.'[56]

1950. From Detroit, a postcard addressed to 'Jesus "Albert" Einstein', sent by a nun.[57]

From London: 'I take the liberty to inform you that Mr George Bernard Shaw is the forger of the Protocols of Zion.'[58]

From Maplewood, New Jersey. 'I read the recent *Life* article [by Lincoln Barnett, 9 January 1950] about your new theory. This is to inform you that "single harmonious edifice of cosmic laws" in which the "deep underlying edifice of the universe is laid bare" happens to be my property. My copyright certificate is dated August 4, 1946.'[59]

From New York: 'Is there any way that I might study under your training . . . I am divorced and am trying to be something my children can look up to.'[60]

A telegram from Milwaukee: 'Found counter measure of atomic bomb.'[61]

A clipping from the *Daily News*, 26 January 1950: 'Einstein's New Theory, New Show at Billy Rose's Diamond Horseshoe.'[62]

From Garmisch-Partenkirchen, a request for assistance to find a buyer for the personal pipe of King Ludwig II of Bavaria.[63]

From Spokane, Washington: 'My thirteen brilliant and talking cats have just told me what the fourth dimension is.'[64]

From Kansas City, Missouri: 'The American people are getting pretty sick of your effrontery in sponsoring the Communist line of appeasement . . . there are very few Jews who can be trusted . . . It might be a good thing if you went to Palestine.'[65]

From Cambridge, Massachusetts, a telegram: 'Attempt to murder me failed. It's mean to use my husband as means to stop me. This time he slashed his own wrist as evidence I was supposed to have attacked him. I call it dirty.'[66]

From Brussels, a proof of God's existence.[67]

1951. From Philadelphia: 'When a day, week or any unit of time passes, where does it go?'[68]

1951

DEAR MR. EINSTEIN
I AM A LITTLE GIRL OF SIX.
I SAW YOUR PICTURE IN THE PAPER.
I THINK YOU OUGHT TO HAVE YOUR HAIRCUT, SO YOU CAN LOOK BETTER.
CORDIALLY YOURS,
ANN G. KOCIN.

Letter of a six-year old girl to Albert Einstein demanding that he have a haircut (1951) (Albert Einstein Archives, Hebrew University of Jerusalem.)

Sender's city unmarked: 'If you and Oppenheimer will run for President and Vice President of the United States in 1952 I shall vote for you.'[69]

From Heerlen, Holland: 'I am the true anti-Christ.'[70]

Sender's city unmarked: 'I thought you might appreciate this. My eight year old son asked me: "What did Einstein study?" I answered "The universe." He replied "Nothing else?"'[71]

From Haifa: 'Sublime brother, I am a scientist theosophist, cabalist. I have in hand the key to the keys to the mystery of all mysteries.'[72]

From Zürich: 'Please inform me whether it is necessary to study physics in order to prolong life.'[73]

1952. From Smyrna, Tennessee, from a Presbyterian minister: 'I am working on a sermon in which I wish to use your most famous formula $E=mc^2$ as an analogy, symbol, or illustration of Christian truth . . . Jesus Christ said "I am the light of the world." It seems more than a coincidence that you should have used the symbol c for light – only I wish it had been C.'[74]

From Rio de Janeiro. A Frenchman sends a collection of poems written by him: 'The Einstein muse, cosmic poems'.[75]

From Berne: 'Announce to the 20th century imbeciles that you have told them a big joke, in order to compromise the mathematical superstition which is in fashion since fifty years.'[76] From Kiel, Germany: 'As *Hausfrau* I cannot bear when a famous man like you walks around so uncared for . . . I am a widow, childless, I lack *Hausfrau* obligations. This stimulated me to make my offer [to care for you]. Please do not take this amiss even if this would not suit you.'[77]

From Auckland, New Zealand: 'Do you believe that all the "races" of people sprang from Adam and Eve?'[78]

A boy from Culvert, Indiana, writes:

> My Father and I are going to build a rocket and go to Mars and Venus. We hope you will go too! We want you to go because we need a good scientist and someone who can guide a rocket good.
>
> Do you care if Mary goes too? She is two years old. She is a very nice girl.
>
> Everybody has to pay for his food because we will go broke if we pay.
>
> 'I hope that you have a nice trip if you go.[79]

1953. From Jackson Heights, New York. 'The origin of the earth is solved; there is only one earth.'[80]

From Portland, Oregon. 'I wonder if you could tell me about anybody who has been taken out of the cemetery and brought back to life. I try to keep up with the news regarding bringing people back to life after death.'[81]

From an unmarked place: 'Would you kindly tell me whether Einstein's hair is coarse or fine in texture.'[82]

From an unmarked place: 'To Albert Einstein. World's greatest faker, racketeer, ego idiot, no mathematician. Chased out of Germany for spreading bullshit that we do not know anything about. RESIGN and go to Palestine. Do you know what KIKES are?'[83]

From Santiago Tianguistenco, Mexico, a request for used stamps: 'We Religious do not collect stamps for the fun of it but we do collect because with the little money we are able to help out at least some of the many thousands and thousands of poor old men and little children to whom poverty and misery have come to reduce to a condition of life unworthy of human beings.'[84]

From the Mother Superior of a convent in Paluvayi, Malabar, India, with an endorsement by the Bishop of Trichur: 'We are in need of a chapel. To build the chapel we require 5000 dollars. We have the foundation stone laid. We have not hitherto put another stone on the foundation stone so now we appeal to your help. In return we have nothing to give you but our humble prayers.'[85]

From the headmaster of the Hospital Corps School, Great Lakes, Illinois: 'Lately I have become extremely concerned about the atheistic convictions of our college students . . . I have read that with all your scientific knowledge you still believe in God . . . I think [you are] a bit worshipped in the world of science . . . if you would tell the people why and how you believe in God it would have such a powerful effect on them.'[86]

From Plainfield, New Jersey. 'I am past 25 years. As yet I have not found a young man my own age to rival your position in my affections. Would you mind temporarily holding this position?'[87]

A telegram from Los Angeles. 'Communism is a mental disease curable within 30 days by an all-out well-planned mental attack which must be utterly sincere, instantaneous, and world-wide to succeed.'[88]

Also from Los Angeles: 'World War III can still be stopped! You are still alive! Mankind has not yet madly risen against the "Inventors of atom bombs". So it is not yet too late for Einstein's atomic-world-peace-ultimatum.'[89] Einstein replied: 'Overburdened with work of all kinds I cannot think of engaging myself in projects like yours.'[90]

A man from Stuttgart sends Einstein some of his poems.[91] Appended is a remark by Einstein: '*Was sich hintern reimt ist darum noch keine Poesie*' ('What ends in rhyme is therefore not yet poetry').

From Dawson, Pennsylvania. 'I wondered why a man so brilliant as you are in so many fields would want to play the violin.'[92]

1954. From Saulte Ste. Marie, Michigan. 'An acquaintance of mine has established himself as a king in his home. He has made his wife a servant.

'His premise is that no great man must do such menial work. Yes, he fancies himself to be a great philosopher. Perhaps, he is.

'I would like to show him a picture of you helping Mrs. Einstein with the dishes. I would be deeply grateful if you could supply one.'[93]

From New York. 'You are a new citizen. Is it courteous for you habitually to criticize your new hosts? If you believe Israel is the place for teachers why don't you go there in a hurry? Please shut up (i.e. stop knocking America and Americans) or renounce your citizenship and serve Israel.'[94]

From Istanbul, a letter addressed 'to him who has made the earth and the heavens tremble, to the founder of the atomic age, Monsieur Albert Einstein, pillar of Israel.'[95]

From Santiago, Chile. 'I have read about your concern for the fate of humanity (in regard to the H-bomb). I am happy to inform you that after years of study I have succeeded to work out a project that in only a few years will abolish the current disorder in the world.'[96]

From Santa Monica, California. 'There are two simple questions which you can no doubt answer: Was Samson's strength in his hair or was it a "gilt"? Also, where exactly did yesterday go?'[97]

The magazine *The Reporter* of 18 November 1954 published Einstein's famous plumber statement:

'. . . If I would be a young man again and had to decide how to make my living, I would not try to become a scientist or scholar or teacher. I would rather choose to be a plumber or a peddlar in the hope to find that modest degree of independence still available under present circumstances . . .'

Among the responses was the following letter from Belle Harbor, New York: 'I noticed in the paper that if you had your ways over again you would want to be a plumber or pedeler. Fortunitly when I was young I used to cell 5¢ socks. Today I am in the whole sale business and sell stretchie socks that can fit anyone from size 9–14. They cost $7 per dozen. If you are interested in becoming a pedeler I will be glad to send some samples on your request.'[98]

1955. On 26 March 1955 Einstein sent this letter to the fifth grade of Farmingdale (N.Y.) Elementary School, thanking the children for a birthday gift:

'Your gift will be an appropriate suggestion to be a little more elegant in the future than hitherto. Because neckties and cuffs exist for me only as remote memories . . .'[79]

Three weeks later, on 18 April at 1.15 in the morning, Einstein died in Princeton Hospital.

References

References to the Einstein Archive give reel number and folder number (in parentheses) followed by document number. See also p. 86 of the text.

1. These are all in (31, 8). 2. (31, 9) 821. 3. (31, 9) 820.
4. (31, 9) 816. 5. (31, 9) 823. 6. (31, 9) 840. 7. (31, 9) 818.
8. (31, 10) 852. 9. (31, 10) 844. 10. (31, 10) 861. 11. (31, 10) 862.
12. (31, 10) 863. 13. (31, 10) 865, 866. 14. (31, 10) 874, 877.
15. (31, 10) 872. 16. (31, 10) 890. 17. (31, 10) 885. 18. (31, 12) 1012.
19. (31, 11) 898. 20. (31, 11) 907. 21. (31, 11) 910. 22. (31, 11) 916.
23. (31, 11) 927.
24. See e.g. the editorial in the *Brooklyn Tablet*, 14 May 1938.
25. (31, 11) 940. 26. (31, 11) 944. 27. (31, 11) 955. 28. (31, 11) 956.
29. (31, 11) 959. 30. (31, 11) 962. 31. (31, 12) 971. 32. (31, 12) 974.
33. (31, 12) 992. 34. (31, 12) 1001. 35. (31, 12) 1007. 36. (31, 12) 1015.
37. (31, 12) 1018. 38. (31, 12) 1025; also (32, 1) 048. 39. (31, 12) 1029.
40. (31, 12) 1032. 41. (31, 12) 1033. 42. (31, 13) 1041. 43. (31, 13) 1047.
44. (31, 13) 1049. 45. (32, 1) 015. 46. (32, 1) 018. 47. (32, 1) 029.
48. (32, 1) 030. 49. (32, 1) 047. 50. (32, 1) 061. 51. (32, 1) 071.
52. (32, 1) 079. 53. (32, 1) 080. 54. (32, 1) 086. 55. (32, 1) 111.
56. (32, 1) 120. 57. (32, 2) 126. 58. (32, 2) 135. 59. (32, 2) 137.
60. (32, 2) 139. 61. (32, 2) 141. 62. (32, 2) 143. 63. (32, 2) 152.
64. (32, 2) 154. 65. (32, 2) 156. 66. (32, 2) 164. 67. (32, 2) 170.
68. (32, 2) 181. 69. (32, 2) 188. 70. (32, 2) 191. 71. (32, 2) 195.
72. (32, 2) 198. 73. (32, 2) 202. 74. (32, 2) 208.
75. (32, 2) 211, 212. 76. (32, 2) 214.
77. (32, 2) 219. 78. (32, 2) 226.
79. This letter from 1952, *not* found in the *komische Mappe*, was shown at an exhibition 'Reflections from Einstein's Archives', at the Jewish National and University Library, Jerusalem, 23 April–3 May 1990. I found this a fitting place to reproduce it.
80. (32, 3) 233. 81. (32, 3) 243. 82. (32, 3) 246. 83. (32, 3) 247.
84. (32, 3) 254. 85. (32, 3) 258. 86. (32, 3) 268. 87. (32, 3) 278.
88. (32, 3) 284. 89. (32, 3) 287. 90. (32, 3) 289. 91. (32, 3) 293.
92. (32, 3) 296. 93. (32, 3) 298. 94. (32, 3) 300. 95. (32, 3) 302.
96. (32, 3) 304. 97. (32, 3) 305. 98. (32, 3) 323.

9

The Indian connection: Tagore and Gandhi

1. Introducing Tagore and Gandhi

Einstein used to refer jokingly to Rabindranath Tagore as Rabbi Tagore.[1] The two men encountered each other for the first time in Germany shortly after the First World War. We have Tagore's reminiscences of their discussions on that occasion.[2] On 14 July 1930 they met again at Einstein's summer house in Caputh near Berlin, where they held a philosophical colloquy that has been published both in the United States[3] and in India.[4] They saw each other briefly for a third – and as far as I know, last – time on 14 December 1930 during Einstein's hectic one-week visit to New York City.[5] There is no record of that meeting. From aboard ship Einstein sent the following telegram to Tagore: 'I congratulate you from my heart with your meeting. May it be given to Tagore also on this occasion to work successfully in the service of his ideal of bringing nations together.'*

Einstein and Gandhi never met face to face. Their entire correspondence consists of one letter by Einstein to Gandhi and one letter of reply. However, Einstein frequently referred with great respect, though not always uncritically, to Gandhi's ideas, as I will show.

I have selected these two men from among the many prominent figures with whom Einstein has been in personal contact or in correspondence because they interest me particularly. They illustrate how widespread some degree of familiarity with Einstein's thinking was. In spite of the different life focus of the three, one the scientist, one the philosopher–artist, one the political activist, they shared profound concerns about the fates of the downtrodden. It is therefore of interest, it seems to me, to take note of Einstein's responses to them, and theirs to him.

* Copy in Einstein Archive. Since the document is undated I cannot tell whether this message was sent before or after their encounter in New York. Nor do I know what meeting is referred to in the telegram.

As a help to readers less familiar with those Indian sages, I shall next briefly introduce them.

Rabindranath Tagore came from a Brahmin background. He was born on 8 May 1861 in Calcutta in Bengal, the oldest Anglo-Indian province. He was the youngest of seven brothers and three sisters. His father, the Maharishi (Great Sage) Debendranath Tagore, was a leader of a religious sect called *Brahma Samaj* (Society of God) which strove to commingle the ancient Hindu traditions with the doctrines of the Christian faith. Rabindranath was to become a revered master within that group. His mother died when he was still a boy. His father's father was an Indian prince.

Rabindranath grew up in a home where music, literature, and drama belonged to life's daily accompaniments. About his school days he has recalled 'The misadventure when I began my career as a student in a school. It was a terribly miserable life which became absolutely intolerable to me . . . The teachers were like living gramophones, repeating the same lessons day by day in a most dull manner. My mind refused to accept anything from my teachers. With all my heart and soul I seem to have repudiated all that was put before me. And then there were some teachers who were utterly unsympathetic and did not understand at all the sensitive soul of a young boy and tried to punish him for the mistakes he made. Such teachers in their stupidity did not know how to teach, how to impart education to a living mind. And because they failed, they punished their victim.'[4] Fortunately, his father allowed him to read by himself or with tutors.

In late adolescence Tagore wrote *Evening songs* and *Morning songs*, two volumes of lyrical poetry in Bengali, which were published before he reached the age of twenty. He married soon afterward.

When he was 23, his father sent him to the country, to manage the family estate on the banks of the Ganges. There he mingled with the village people and began writing about their daily life.

At the age of forty he established in the Calcutta region the *Shantiniketan*, which means 'abode of peace', helped with funds provided by his father. It was a school devoted to the study of the whole range of Eastern culture. He now engaged in teaching and in the building up of his school, continuing writing, and began to translate some of his works into English.

In 1912 Tagore went abroad, visiting England and spending the winter of 1912–13 in the U.S.A. His lectures before university audiences, given in both countries, have been collected in his book *Sadhana, the realization of life*.

Meanwhile, translations in English of some of Tagore's writings had begun to appear, first his *Gitánjali* (song offerings), with a preface by

William Butler Yeats, then his 'Glimpses of Bengal Life'. These publications caused him to be hailed as a new master, 'combining at once the feminine grace of poetry with the virile power of prose'.[6] By 1913 he had furthermore published in English a cycle of poems describing the bliss and torture of young love; and a collection of children's poems.

In October 1913, a reader of *Gitánjali* commented as follows: 'This small collection of poems, translated into English by the author himself . . . makes such a surprisingly rich and truly poetic impression, that there is nothing baroque in proposing to award it even with a distinction such as here is in question' (the Nobel Prize for literature).* Further in this report we read: 'At issue here is undoubtedly a more remarkable revelation than any European poetry has to offer at this time . . . [Perhaps] it goes against the grain to award this religious poetry with a prize. It is as if one were to pay for David's psalms or the songs of St Francis . . . [Tagore] is recognized as his country's master of Scaldic art; no European bard can presently make such claims.'

In December 1913, the prize was awarded to Tagore 'for the profound and noble inspiration, for the beauty and novelty which his poetic genius has introduced in brilliant fashion in Western literature'.[6] Being unable to receive his prize in person, he sent the following telegram to Stockholm: 'I beg to convey [to] the Swedish Academy my grateful appreciation of the breadth of understanding, which has brought the distant near and has made a stranger a brother.'[6] He used his prize money for the upkeep of *Shantiniketan*.

In his later years Tagore continued to publish volumes of poems, a few novels, several plays, and volumes of short stories. He also became a composer, setting some hundred Indian poems to music, and was one of India's foremost painters of his day.

Mohandas Karamchand Gandhi, born in Porbandar in 1869, came to be called in later life the *Mahatma*, the Great Souled. His father was a high administrator in the Anglo–Indian government. His upbringing at home taught him *ahimsa*, the principle of non-injury to all living beings. He married at the age of 13 and had four sons. Gandhi was sent to London to study law, which he briefly practiced in the Bombay High Court.

In 1893, professional business brought him to South Africa, where he stayed for 21 years. While traveling in a first-class railway compartment in Natal he was asked by a white man to leave, which he did. That incident

* These lines are found in the report on Tagore's *oeuvre*, addressed to the *Svenska Akademiens Nobelkommittee*. This Nobel Prize proposal was written by the Swedish author Per August Hallström. I am much indebted to *Svenska Akademiens Arkivet* for permission to quote from this document.

marked a watershed in his life, eventually causing him to begin his political activities. He threw himself into the struggle for liberties for Indian settlers, campaigning for *Satyagraha*, insistence on truth, and against colonialism, racism, and violence. His strategy of non-violent disobedience to suppressive laws was new and may be called revolutionary.

By 1920 he had become a dominant figure in Indian politics. His persistence against heavy odds caused him to be imprisoned for several protracted periods.

He was a major force in the struggle for independence of India, which was granted in 1947. He lived to see that event, but not for long. On 30 January 1948 he was shot and killed by Nathuram Godse, a young Indian fanatic.

The collected edition of Gandhi's writings runs to more than eighty volumes. His concepts have inspired such later fighters for civil liberties as Martin Luther King.

2. Einstein and Tagore

Tagore has left us these personal impressions of Einstein:

> Einstein has often been called a lonely man. Insofar as the realm of the mathematical vision helps to liberate the mind from the crowded trivialities of daily life, I suppose he is a lonely man. His is what might be called a transcendental materialism, which reaches the frontiers of metaphysics, where there can be utter detachment from the entangling world of self. To me both science and art are expressions of our spiritual nature, above our biological necessities and possessed of an ultimate value.
>
> Einstein is an excellent interrogator. We talked long and earnestly about my 'religion of man'. He punctuated my thoughts with terse remarks of his own, and by his questions I could measure the trend of his own thinking.[2]

I turn to the main points of the first Einstein–Tagore conversation.*

E. 'Do you believe in the Divine as isolated from the world?'

T. 'Not isolated. The infinite personality of Man comprehends the Universe. There cannot be anything that cannot be subsumed by the human personality, and this proves that the truth of the Universe is human truth.'

E. 'There are two different conceptions about the nature of the Universe: (1) the world as a unity dependent on humanity; (2) the world as a reality independent of the human factor.'

* As recorded in ref. 2. In ref. 4 these are stated to have been made during their second dialogue. It is my impression that the correct assignment is as in ref. 2.

T. 'When our universe is in harmony with man, the eternal, we know it as truth, we feel it as beauty.'

E. 'This is the purely human conception of the universe.'

T. 'Yes, one eternal entity. We have to realize it through our emotions, and activities. We realized the Supreme Man who has no individual limitations through our limitations. Science is concerned with that which is not confined to individuals, it is the impersonal human world of truths. Religion realizes these truths and links them up with our deeper needs; our individual consciousness of truth gains universal significance. Religion applies values to truth, and we know truth as good through our own harmony with it.'

E. 'Truth, then, or Beauty is not independent of Man?'

T. 'No.'

E. 'I agree with regard to this conception of Beauty, but not with regard to Truth.'

T. 'Why not? Truth is realized through man.'

E. 'I cannot prove that scientific truth must be conceived as a truth that is valid independent of humanity; but I believe it firmly. I believe, for instance, that the Pythagorean theorem in geometry states something that is approximately true, independent of the existence of man. Anyway, if there is a *reality* independent of man, there is also a truth relative to this reality; and in the same way the negation of the first engenders a negation of the existence of the latter.'

T. 'Truth, which is one with the Universal Being, must essentially be human; otherwise whatever we individuals realize as true can never be called truth, at least the truth which is described as scientific and which only can be reached through the process of logic, in other words, by an organ of thoughts which is human . . . The nature of truth which we are discussing is an appearance, that is to say, what appears to be true to the human mind and therefore is human, and may be called *Maya* or illusion.'

E. 'The problem begins whether Truth is independent of our consciousness.'

T. 'What we call truth lies in the rational harmony between the subjective and objective aspects of reality, both of which belong to the superpersonal man.'

E. 'Even in our everyday life, we feel compelled to ascribe a reality independent of man to the objects we use. We do this to connect the experiences of our senses in a reasonable way. For instance, if nobody is in this house, yet that table remains where it is.'

T. 'Yes, it remains outside the individual mind but not the universal mind. The table which I perceive is perceptible by the same kind of consciousness which I possess.'

E. 'Our natural point of view in regard to the existence of truth apart from humanity cannot be explained or proved, but it is a belief which nobody can lack – no primitive beings even. We attribute to Truth a superhuman objectivity, it is indispensable for us, this reality which is independent of our existence and our experience and our mind – though we cannot say what it means.'

T. 'In the apprehension of truth there is an eternal conflict between the universal human mind and the same mind confined in the individual. The perpetual process of reconciliation is being carried on in our science, philosophy, in our ethics. In any case, if there be any truth absolutely unrelated to humanity, then for us it is absolutely non-existing . . . If there be some truth which has no sensuous or rational relation to human mind, it will ever remain as nothing so long as we remain human beings.'

E. 'Then I am more religious than you are!'

T. 'My religion is in the reconciliation of the Super-personal Man, the Universal human spirit, in my own individual being . . . In science we go through the discipline of eliminating the personal limitations of our indi-

Einstein with Rabindranath Tagore in Caputh, July 14, 1930. (AIP Emilio Segrè Visual Archives.)

vidual minds and thus reach that comprehension of truth which is in the mind of the Universal Man.'

Later Tagore summarized this discussion as follows: 'Einstein desired to know whether I held that the Divine was isolated from the world. To me, the great spiritual man is a fact not wholly fathomed but existent and being realized from within and from without . . . Man's steadying pivot is his belief that truth beckons him, even when he does not see it. This is the call to the individual from the universal man, and it is this call that carries us toward the universal. Animals are happy, hearing it not. Man is not happy; yet he is eager for the adventure.

"So this is the realization of the human entity?" asked Einstein. He grasped my view of the universal truth of existence not as an abstraction but as a reality spiritually related to human personality.

I could see that Einstein held fast to the extra-human aspect of truth. But it is evident to me that in human reason facts assume a unity of truth which is only possible to a human mind.'[2]

Einstein later invented the term 'Objective reality'[7] for his deep belief in the extra-human nature of truth. Tagore could not accept this view. Neither can nearly all modern physicists, but for quite different reasons. This issue also was central to the dialogue between Einstein and Niels Bohr[8] who, like most of us, could not accept Einstein's opinion – not on philosophical but on physical grounds.

The second Einstein–Tagore dialogue,* the one held in Caputh, 'was taken down by a friend who was present'.[2] This time they began with a discussion of the nature of causality. On this subject the two men talked past each other without any understanding as to what the other was driving at. I do not consider it worth while reproducing here any of this. On the other hand, their next theme, on music is quite appealing.

T. 'The musical system in India . . . is not so rigidly fixed as is the western music. Our composers give a certain definite outline, a system of melody and rhythmic arrangement, and within a certain limit the player can improvise upon it. He must be one with the law of that particular melody, and then he can give spontaneous expression to his musical feeling within the prescribed regulation. We praise the composer for his genius in creating a foundation along with a superstructure of melodies, but we expect from the player his own skill in the creation of variations of melodic flourish and ornamentation. In creation we follow the central law of existence, but, if

* Prior to that event, on 22 December 1929, Tagore had sent a postcard to Einstein from India: 'My salutation is to him who knows me imperfect and loves me. My best wishes.'[9] I do not know what was the occasion for sending this note.

we do not cut ourselves adrift from it, we can have sufficient freedom within the limits of our personality for the fullest self-expression.'

E. 'That is only possible where there is a strong artistic tradition in music to guide the people's mind. In Europe, music has come too far away from popular art and popular feeling and has become something like a secret art with conventions and traditions of its own.'

T. 'So you have to be absolutely obedient to this too complicated music. In India the measure of a singer's freedom is in his own creative personality. He can sing the composer's song as his own, if he has the power creatively to assert himself in his interpretation of the general law of melody which he is given to interpret.'

E. 'It requires a very high standard of art fully to realize the great idea in the original music, so that one can make variations upon it. In our country the variations are often prescribed.'

T. 'If in our conduct we can follow the law of goodness, we can have real liberty of self-expression. The principle of conduct is there, but the character which makes it true and individual is our own creation. In our music there is a duality of freedom and prescribed order.'

E. 'Are the words of a song also free? I mean to say, is the singer at liberty to add his own words to the song which he is singing?'

T. 'In Bengal we have a kind of song – *Kirtan*, we call it – which gives freedom to the singer to introduce parenthetical comments, phrases not in the original song. This occasions great enthusiasm, since the audience is constantly thrilled by some beautiful, spontaneous sentiment freshly added by the singer.'

E. 'Is the metrical form quite severe?'

T. 'Yes, quite. You cannot exceed the limits of versification; the singer in all his variations must keep the rhythm and the time, which is fixed. In European music you have a comparative liberty about time, but not about melody. But in India we have freedom of melody with no freedom of time.'

E. 'Can the Indian music be sung without words? Can one understand a song without words?'

T. 'Yes, we have songs with unmeaning words, sounds which just help to act as carriers of the notes. In North India music is an independent art, not the interpretation of words and thoughts, as in Bengal. The music is very intricate and subtle and is a complete world of melody by itself.'

E. 'It is not polyphonic?'

T. 'Instruments are used, not for harmony, but for keeping time and for adding to the volume and depth. Has melody suffered in your music by the imposition of harmony?'

E. 'Sometimes it does suffer very much. Sometimes the harmony swallows up the melody altogether.'

T. 'Melody and harmony are like lines and colors in pictures. A simple linear picture may be completely beautiful; the introduction of color may make it vague and insignificant. Yet color may, by combination with lines, create great pictures, so long as it does not smother and destroy their value.'

E. 'It is a beautiful comparison; line is also much older than color. It seems that your melody is much richer in structure than ours. Japanese music seems to be so.'

T. 'It is difficult to analyze the effect of eastern and western music on our minds. I am deeply moved by the western music – I feel that it is great, that it is vast in its structure and grand in its composition. Our own music touches me more deeply by its fundamental lyrical appeal. European music is epic in character; it has a broad background and is Gothic in its structure.'

E. 'Yes, yes, that is very true. When did you first hear European music?'

T. 'At seventeen, when I first came to Europe. I came to know it intimately, but even before that time I had heard European music in our own household. I had heard the music of Chopin and others at an early age.'

E. 'There is a question we Europeans cannot properly answer, we are so used to our own music. We want to know whether our own music is a conventional or a fundamental human feeling, whether to feel consonance and dissonance is natural or a convention which we accept.'

T. 'Somehow the piano confounds me. The violin pleases me much more.'

E. 'It would be interesting to study the effects of European music on an Indian who had never heard it when he was young.'

T. 'Once I asked an English musician to analyze for me some classical music and explain to me what are the elements that make for the beauty of a piece.'

E. 'The difficulty is that the really good music, whether of the East or of the West, cannot be analyzed.'

T. 'Yes, and what deeply affects the hearer is beyond himself.'

E. 'The same uncertainty will always be there about everything fundamental in our experience, in our reaction to art, whether in Europe or Asia. Even the red flower I see before me on your table may not be the same to you and me.'

T. 'And yet there is always going on the process of reconciliation between them, the individual taste conforming to the universal standard.'[2]

We know of Einstein's reactions to his meetings with Tagore from the reply he sent on 10 October 1930 to a letter from the French author and pacifist

Romain Rolland, soliciting a contribution to a book to be presented to Tagore on the occasion of his seventieth birthday, in May 1931:

> I shall be glad to sign your beautiful text and to add a brief contribution. My conversation with Tagore was rather unsuccessful because of difficulties in communication and should, of course, never have been published. In my contribution, I should like to give expression to my conviction that men who enjoy the reputation of great intellectual achievement have an obligation to lend moral support to the principle of unconditional refusal of war service . . .[10]*

The 'beautiful text' presumably refers to the somewhat exalted preface to *The Golden Book of Tagore*, which came out in 1931. Part of this preface, signed by Einstein, Gandhi, and Rolland, reads:

> He has been for us the living symbol of the Spirit, of Light, and of Harmony – the great free bird which soars in the midst of tempests – the song of Eternity which Ariel strikes on his golden harp, rising above the sea of unloosened passions. But his art never remained indifferent to human misery and struggles. He is the 'Great Sentinel'.
>
> For all that we are and we have created have had their roots and their branches in that Great Ganges of Poetry and Love.

Einstein's contribution to the book reads:

> You are aware of the struggle of creatures that spring forth out of need and dark desires. You seek salvation in quiet contemplation and in the workings of beauty. Nursing these you have served mankind by a long, fruitful life, spreading a mild spirit, as has been proclaimed by the wise men of your people.[9]

In May 1931, Tagore sent Einstein a postcard, written in Bengali and in English, thanking him for his tribute: 'The same sun is newly born in new lands, in a ring of endless dawns.'[9]

A final word. In 1932 a Persian mathematician asked Tagore what were his opinions of Einstein. 'Tagore, cheery and smiling, said: "I shall not go into his mathematics and his science. He is a good and friendly man who has withdrawn himself from the world and its superficialities. He is devoted to peace and the wellbeing of mankind. He has pledged almost his entire life to those ideas. In his speeches in America he has stressed the harm of war and the benefits of peace. He is one of the greatest thinkers of our time. In my opinion he is no way fanatic about race. He looks upon all people as equal and is fully prepared to serve humanity.'[12]

* Among the signatories of a manifesto released on 12 October 1930, an appeal against conscription and the military training of youth, we find the names of Einstein, Tagore, and Rolland.[11]

3. Einstein and Gandhi

Browsing through Einstein's writings, I found expressions of his admiration for Gandhi as early as 1929[13] and as late as 1954. The *Mahatma* clearly had made a much deeper impression on Einstein than had Tagore.

The exchange of letters between the two men, mentioned earlier, occurred in the autumn of 1931. Einstein wrote first. I do not know for certain why he did so at that time. My best guess is that in Berlin he had met a Mr Sundaram, an acquaintance of Gandhi, who had taken his letter to London, where Gandhi was attending the Round Table Conference on Indian Constitutional Reform. Here is what Einstein wrote:

> You have shown by all you have done that we can achieve the ideal even without resorting to violence. We can conquer those votaries of violence by the non-violent method. Your example will inspire and keep humanity to put an end to a conflict based on violence with international help and cooperation guaranteeing peace of the world. With this expression of my devotion and admiration I hope to be able to meet you face to face.[14]

On 10 October 1931 Gandhi replied from London:

> I was delighted to have your beautiful letter sent through Sundaram. It is a great consolation for me that the work I am doing finds favor in your sight. I do indeed wish that we could meet face to face and that too in India at my Ashram.[9]

The only other statement by Gandhi present in the Einstein Archive is an undated statement, in German, about Zionism. I did not even find a direct indication that it was addressed to Einstein. It is an interesting comment, though, which I give next in English translation:

> The Zionism in its spiritual sense is a noble aspiration, but the Zionism which aims at the re-occupation of Palestine by Jews does not appeal to me. I understand the yearning of the Jew to return to the land of his forefathers. He can and should do that in so far as this return can be achieved without English or Jewish bayonets. In that case the Jew who goes to Palestine can live in perfect peace and friendship with the Arabs. The real Zionism which rests in the hearts of the Jews is an aim for which one should strive and give one's life. Such a Zionism is the abode of God. The true Jerusalem is a spiritual Jerusalem. And that spiritual Zionism can be realized by the Jew in every part of the world.

The remainder of this section presents comments by Einstein on Gandhi's ideas.

1932. 'While Gandhi's economic views are questionable, his *Satyagraha* is very important, and could be applied to Europe's problems.'[15]

1935. Einstein is specific in regard to Gandhi's economic proposals.

> I admire Gandhi greatly but I believe there are two weaknesses in his program; while nonresistance is the most intelligent way to cope with adversity, it can be practiced only under ideal conditions. It may be feasible to practice it in India against the British but it could not be used against the Nazis in Germany today. Then, Gandhi is mistaken in trying to eliminate or minimize machine production in modern civilization. It is here to stay and must be accepted.[16]

1939. Einstein receives a letter[17] from one of Gandhi's followers requesting a contribution to a volume of essays to be presented to Gandhi on his seventieth birthday. Einstein did so:

> *Mahatma* Gandhi's life's work is unique in political history. He has devised a quite new and humane method for fostering the struggle for liberation of his suppressed people and has implemented it with greatest energy and devotion. The moral influence which it has exerted on the consciously thinking people of the entire civilized world might be far more lasting than may appear in our time of overestimation of brutal methods of force. For only the work of such statesmen is lasting who by example and educational action awaken and establish the moral forces of their people.
>
> We may all be happy and grateful that fate has given us such a shining contemporary, an example for coming generations.[18]

1950. We are now in the years after India's independence. Nehru is prime minister of India and Gandhi has been murdered.

'. . . I have studied the works of Gandhi and Nehru with real admiration. India's forceful policy of neutrality in regard to the American–Russian conflict could well lead to a unified attempt on the part of the neutral nations to find a supranational solution to the peace problem.'[19]

Also in 1950, during a broadcast sponsored by the United Nations under the title 'The pursuit of peace', Einstein declares: 'I believe that Gandhi held the most enlightened views of all the political men in our time. We should strive to do things in his spirit; not to use violence in fighting for our cause and to refrain from taking part in anything we believe is evil.'[20]

1951. Einstein in a letter: 'Revolution without the use of violence was the method by which Gandhi brought about the liberation of India. It is my belief that the problem of bringing peace to the world on a supranational basis will be solved only by employing Gandhi's method on a large scale.'[21]

1952. In a letter to the Asian Congress for World Federation, held in Hiroshima, 3–6 November: 'There is little point . . . in opposing the manufacture of *specific* weapons [nuclear bombs]; the only solution is to abolish both war and the threat of war . . .'

'Gandhi, the greatest political genius of our time, indicated the path to be taken. He gave proof of what sacrifice man is capable once he has discovered the right path. His work in behalf of India's liberation is living testimony to the fact that man's will, sustained by an indomitable conviction, is more powerful than material forces that seem insurmountable.'[22]

1953. In another letter: 'It should not be forgotten that Gandhi's development resulted from extraordinary intellectual and moral forces in combination with political ingenuity and a unique situation.'[23]

1954. And in yet another letter: 'I express openly what I think. But I know that does not mean that I could create a popular movement such as Gandhi was able to do. You can be sure that nothing can be achieved solely by preaching reason.'[24]

On numerous occasions Einstein wrote eulogies, all of substance, on a great variety of people. It is fitting to conclude this essay with his brief but masterful words prepared for the memorial service for Gandhi held in Washington, D.C., on 11 February 1948 – three weeks after his murder:

> Everyone concerned with a better future for mankind must be deeply moved by the tragic death of Gandhi. He died a victim of his own principle, the principle of nonviolence. He died because, in a time of disorder and general unrest in his country, he refused any personal armed protection. It was his unshakable belief that the use of force is an evil in itself, to be shunned by those who strive for absolute justice.
>
> To this faith he devoted his whole life, and with this faith in his heart and mind he led a great nation to its liberation. He demonstrated that the allegiance of men can be won, not merely by the cunning game of political fraud and trickery, but through the living example of a morally exalted way of life.
>
> The veneration in which Gandhi has been held throughout the world rests on the recognition, for the most part unconscious, that in our age of moral decay he was the only statesman who represented that higher conception of human relations in the political sphere to which we must aspire with all our powers. We must learn the difficult lesson that the future of mankind will only be tolerable when our course, in world affairs as in all other matters, is based upon justice and law rather than the threat of naked power, as has been true so far.[25]

References

1. Private communication from the late Helen Dukas.
2. R. Tagore, in *Asia Magazine*, **31**, 139, 1931.
3. *The American Hebrew*, 11 September 1931, p. 351.

4. *The Modern Review*, Calcutta, January 1931, p. 42.

5. See 'Einstein and the press', Chapter 11, Section 5.

6. *Les Prix Nobel*, Norstedt and Sons, Stockholm, 1914.

7. A. Einstein, B. Podolski, and N. Rosen, *Phys. Rev.*, **47**, 777, 1935.

8. For details see *SL*, Chapter 25, Section (c).

9. Copy in Einstein Archive.

10. O. Nathan and H. Norden, *Einstein on peace*, p. 112, Schocken, New York, 1968.

11. Ref. 10, p. 113.

12. S.D. Tehranir, in the Persian journal *Armaghan*, 1932. Undated copy in the Einstein Archive.

13. See e.g. ref. 10, p. 98.

14. Copy in Einstein Archive, undated.

15. Interview with *The Friend* (the British Quaker Journal), 12 August 1932.

16. Interview with the *Survey Graphic*, August 1935.

17. S. Radhakrishnan, letter to A. Einstein, 12 January 1939.

18. *Birthday Volume to Gandhi*, Allen and Unwin, London, 1939.

19. Ref. 10, p. 525.

20. *NYT*, 19 June 1950.

21. Ref. 10, p. 543.

22. Ref. 10, p. 584.

23. Ref. 10, p. 594.

24. Ref. 10, p. 606.

25. Ref. 10, p. 467.

10

Einstein on religion and on philosophy

1. A Fiddler on the Roof

You may have seen the musical *Fiddler on the Roof*; if not, you should. It is a tale of happiness and sorrow set in a small Jewish community somewhere in Eastern Europe. As the curtain rises one sees an old Jew sitting on the roof of his small house, playing his fiddle.

Why would he sit just there when playing? I understand his purpose to be that he wishes to be as near as possible to his God in the heavens, thus symbolizing the closeness the orthodox Jew feels to his Lord.

I believe that the relation between the Jew and God is unique in certain respects. He knows that he shall and will obey Him but that will not prevent him from, once in a while, giving Him an argument before doing so.

That opening scene of the musical has often reminded me of Einstein, not just because in actual fact he too did fiddle but, more so, because of the particular ways in which he would invoke God's image in spoken and written language. Let me give some examples.

In 1897, at the age of 18, he wrote a friend: 'Strenuous labor and the contemplation of God's nature are the angels which, reconciling, fortifying, and yet mercilessly severe, will guide me through the tumult of life.'[1]

In the autumn of 1919, in the course of a discussion with a student, Einstein – now aged 40 – handed her a cable which had informed him that the bending of light by the sun was in agreement with his general relativistic prediction. The student asked what he would have said if there had been no confirmation. Einstein replied, '*Da könnt' mir halt der liebe Gott leid tun. Die Theorie stimmt doch.*' 'Then I would have to pity the dear Lord. The theory is correct anyway.'[2]

Two years later, in May 1921, Einstein lectured at Princeton University. While there, word reached him of an experimental result which, if true – it turned out not to be – would contradict his theory. Upon hearing this rumor, Einstein commented: '*Raffiniert ist der Herr Gott, aber boshaft ist Er nicht.*' 'Subtle is the Lord, but malicious He is not.'[3]

In 1942, aged 63, Einstein wrote to a colleague: 'It seems hard to look into God's cards. But I cannot for a moment believe that He plays dice and makes use of "telepathic" means (as the current quantum theory alleges He does).'[4]

These stories show Einstein as the instrument of the Lord – trying to look into His cards and prepared to give Him an occasional argument.

Do these tales indicate that Einstein was religious? Yes and no, as I shall attempt to explain. To introduce this subject let us see how Einstein responded to religion from boyhood on.

2. Upbringing

On 8 August 1876, Hermann Einstein and Pauline Koch, Einstein's future parents, were married in a synagogue in a small town in Germany. On a sunny Friday in March 1879 their first child was born, a citizen of the new German empire. On the following day Hermann went to register the birth of his son. In translation the birth certificate reads, 'No. 224, Ulm, 15 March 1879. Today, the merchant Hermann Einstein, residing in Ulm, Bahnhofstrasse 135, of the Israelitic faith, personally known, appeared before the undersigned registrar, and stated that a child of the male sex, who has received the name Albert, was born in Ulm, in his residence, to his wife Pauline Einstein, née Koch, of the Israelitic faith, on 14 March of the year 1879, at 11.30 a.m. Read, confirmed, and signed. Hermann Einstein. The Registrar, Hartman.' Albert was named (if one may call it that) after his grandfather Abraham.[5]

'A liberal spirit, nondogmatic in regard to religion, prevailed in the family. Both parents had themselves been raised that way. Religious matters and precepts were not discussed.'[6] Albert's father was proud of the fact that Jewish rites were not practiced in his home.[6]

Shortly after Albert's birth, the family moved to Munich in Bavaria. There, Einstein, at the age of about six, entered elementary school, the *Volksschule*. In October 1888 Albert moved from there to the *Luitpold Gymnasium*, which was to be his school until he was fifteen.

Bavarian law required that all children of school age receive religious education. At the *Volksschule*, only instruction in Catholicism was provided. Einstein was taught the elements of Judaism at home by a distant relative.[7] When he went to the *Luitpold Gymnasium*, this instruction continued at school. As a result of this inculcation, Einstein went through an intense religious phase when he was about eleven years old. His feelings were of such ardor that he followed religious precepts in detail. For example, he ate no pork.[8] Later, in his Berlin days, he told a close friend that during this period he had composed several songs in honor of God, which he sang enthusiastically to himself on his way to school.[9] This interlude came to an abrupt end a year later as a result of his exposure to science. Einstein himself has recalled how profound an impact this abandonment had on his entire later thinking:

> Through the reading of popular scientific books I soon reached the conviction that much in the stories of the Bible could not be true. The consequence was a positively fanatic [orgy of] freethinking coupled with the impression that youth is intentionally being deceived by the state through lies; it was a crushing impression. Suspicion against every kind of authority grew out of this experience, a skeptical attitude towards the convictions which were alive in any specific social environment – an attitude which has never again left me, even though later on, because of a better insight into the causal connections, it lost some of its original poignancy.
>
> It is quite clear to me that the religious paradise of youth, which was thus lost, was a first attempt to free myself from the chains of the 'merely-personal', from an existence which is dominated by wishes, hopes, and primitive feelings. Out yonder there was this huge world, which exists independently of us human beings and which stands before us like a great, eternal riddle, at least partially accessible to our inspection and thinking. The contemplation of this world beckoned like a liberation, and I soon noticed that many a man whom I had learned to esteem and to admire had found inner freedom and security in devoted occupation with it. The mental grasp of this extra-personal world within the frame of the given possibilities swam as highest aim half consciously and half unconsciously before my mind's eye. Similarly motivated men of the present and of the past, as well as the insights which they had achieved, were the friends which could not be lost. The road to this paradise was not as comfortable and alluring as the road to the religious paradise; but it has proved itself as trustworthy, and I have never regretted having chosen it.[10]

Einstein did not become *bar mitzvah*. He never mastered Hebrew. When he was fifty, Einstein wrote to *Oberlehrer* Heinrich Friedmann, his religion teacher at the *Gymnasium*, 'I often read the Bible, but its original text has remained inaccessible to me.'[11] His brief religious ardor had left no trace, just as in later years he would often wax highly enthusiastic about a scientific idea, then drop it as of no consequence.

3. Judaism in Einstein's early career

In 1905 Einstein, now a clerk at the Swiss Patent Office in Berne, came forth with his special theory of relativity, his lightquantum hypothesis, and his theory of Brownian motion. Each of these three contributions would separately have sufficed to establish him for ever as a great twentieth-century figure in science. That was not at once clear, however, as witness the difficulties he experienced during the next few years in obtaining an academic position. The complex of reasons for that need not be gone into here, with the exception of the influence his Jewish descent played.

In 1908 Einstein wrote to a friend, asking him what was the best way to apply for a vacant high school position. 'Can I visit there to give an oral demonstration of my laudable personality as teacher and citizen? Wouldn't I probably make a bad impression (no Swiss-German, Semitic looks, etc.)? Would it make sense if I were to extol my scientific papers on that occasion?'[12] Perhaps he never applied, perhaps he was rejected.

In 1909, Einstein did obtain his first faculty position, however, as associate professor at the University of Zürich. This proposal to the faculty clearly shows his rapidly growing renown. 'Today Einstein ranks among the most important theoretical physicists and has been recognized rather generally as such since his work on the relativity principle . . . uncommonly sharp conception and pursuit of ideas . . . clarity and precision of style . . .'[13]

Einstein must have been aware of this appreciation. Perhaps, also, he may have sensed some of the following sentiments expressed in a part of the final faculty report.*

> These expressions of our colleague, based on several years of personal contact, were all the more valuable for the committee as well as for the faculty as a whole since Herr Dr Einstein is an Israelite and since precisely to the Israelites among scholars are ascribed (in numerous cases not entirely without cause) all kinds of unpleasant peculiarities of character, such as intrusiveness, impudence, and a shopkeeper's mentality† in the perception of their academic position. It should be said, however, that also among the Israelites there exist men who do not exhibit a trace of these disagreeable qualities and that it is not proper, therefore, to disqualify a man only because he happens to be a Jew. Indeed, one occasionally finds people also among non-Jewish scholars who in regard to commercial perception and utilization of their academic profession develop qualities which are usually considered as specifically 'Jewish'. Therefore neither the committee nor the faculty as a whole considered it compatible with its dignity to adopt anti-Semitism as a matter of policy and the information which Herr Kollege Kleiner was able to provide about the character of Herr Dr Einstein has completely reassured us.[14]

Opinions such as these of course do not describe just Zürich in 1909 but western civilization in the early twentieth century.

* It is, of course, highly improbable that Einstein ever saw this report.

† . . . '*Zudringlichkeit, Unverschämtheit, Krämerhaftigkeit* . . .'

The secret faculty vote of March 1909 on the Einstein appointment was ten in favor, one abstention.

One further episode in Einstein's early career bears on Einstein the Jew. In 1910 he received word that efforts were under way to appoint him to a full professorship at the Karl-Ferdinand University in Prague. The procedures took time and in the summer of 1910 Einstein wrote to a colleague: 'I did not receive the call from Prague. I was only proposed by the faculty; the ministry has not accepted the proposal because of my Semitic descent.'[15] (I have seen no documents to that effect.) In January 1911 he did receive the official call, however. Prior to the beginning of his appointment he had to record his religious affiliation. The answer *none* was unacceptable. He wrote '*Mosaisch*'.[16]

Einstein stayed in Prague only until July 1912.[17] Prague newspapers expressed regrets about his departure. One of them added this interesting comment: 'In the German empire Jews cannot be judges or lieutenants, but in the chairs at German universities they enjoy fully equal rights. The number of Israelitic full professors in Germany is at present about twenty-five.'[18] A few days later Einstein said in an interview with a Viennese paper: 'In Prague I did not notice any denominational prejudice, as others have surmised.'[19]

In 1914 Einstein accepted a position in Berlin, where anti-semitism hit him hard, as has been noted elsewhere in this volume.[20]

4. 'A religious feeling of a special kind'

In his mature years Einstein often thought, spoke, and wrote about his conception of religion. We have seen that as a boy he had freed himself from the orthodoxy of church and synagogue. So it remained, as he stated clearly in 1936 (at the age of 57) in reply to a letter from a young girl who had written him to ask whether scientists pray and, if so, what they pray for. From his reply:

> A scientist will hardly be inclined to believe that the course of events can be influenced by prayer – that is by a wish addressed to a supernatural being . . .
>
> On the other hand, everyone seriously engaged in science reaches the conviction that the Laws of Nature manifest a spirit which is vastly superior to Man, and before which we, with our modest strength, must humbly bow. Hence does the preoccupation with science lead to a religious feeling of a specific kind which, however, differs essentially from the religiosity of more naïve people.[21]

This letter contains the reasons why I answered my own question whether Einstein was religious with 'yes and no': 'yes' in the special sense just mentioned; 'no' not in the naïve sense.

The same emphasis on both yes and no is found in Einstein's article of 1930 in which he called himself a deeply religious man:

> A knowledge of the existence of something we cannot penetrate, our perceptions of the profoundest reason and the most radiant beauty, which only in their most primitive forms are accessible to our minds – it is this knowledge and this emotion that constitute true religiosity; in this sense, and in this alone, I am a deeply religious man. I cannot conceive of a God who rewards and punishes his creatures, or has a will of the kind that we experience in ourselves. Neither can I nor would I want to conceive of an individual that survives his physical death; let feeble souls, from fear or absurd egoism, cherish such thoughts. I am satisfied with the mystery of the eternity of life and with the awareness and a glimpse of the marvelous structure of the existing world, together with the devoted striving to comprehend a portion, be it ever so tiny, of the Reason that manifests itself in nature.[22,23]

In that same year Einstein wrote another article, which I find very interesting, in which he speculated that religious notions in general and the formation of the church in particular has originally arisen out of fear:

> With primitive man it is above all fear that evokes religious notions – fear of hunger, wild beasts, sickness, death. Since at this stage of existence understanding of causal connections is usually poorly developed, the human mind creates illusory beings more or less analogous to itself on whose wills and actions these fearful happenings depend. Thus one tries to secure the favor of these beings by carrying out actions and offering sacrifices which, according to the tradition handed down from generation to generation, propitiate them or make them well disposed toward a mortal. In this sense I am speaking of a religion of fear. This, though not created, is in an important degree stabilized by the formation of a special priestly caste which sets itself up as a mediator between the people and the beings they fear, and erects a hegemony on this basis. In many cases a leader or ruler or a privileged class whose position rests on other factors combines priestly functions with its secular authority in order to make the latter more secure; or the political rulers and the priestly caste make common cause in their own interests.[24]

Elsewhere Einstein outlined in broad strokes the place of religion in the life of modern man:

> What is the meaning of human life, or, for that matter, of the life of any creature? To know an answer to this question means to be religious. You ask: Does it make any sense, then, to pose this question? I answer: The man who regards his own life and that of his fellow creatures as meaningless is not merely unhappy but hardly fit for life.[25]

It is noteworthy that all these commentaries on religion of Einstein's stem from the 1930s. The same is true for his reflections on the relations between science and religion. I do not know why these preoccupations would all date from that period. Perhaps that had to do with the menace to civilization brought about by the rise of the Nazis.

5. Science and religion

In 1950 Einstein was interviewed in his home in Berlin on the subject of science and God.[26] The first question posed to him was to comment on a recent suggestion made during a meeting of American scientists that the time had come for science to give a new definition of God. Einstein's reply was terse: 'Quite ridiculous.' Asked for his views on the public yearning for science to give spiritual help and inspiration which organized religion no longer seemed able to give, he answered: 'Speaking of the spirit that informs modern scientific investigations, I am of the opinion that all the finer speculations in the realm of science spring from a deep religious feeling, and that without such feeling they would not be fruitful. I also believe that this kind of religiousness which makes itself felt to-day in scientific investigation is the only creative religious activity of our time. The art of to-day can hardly be looked upon at all as expressive of our religious instincts . . . But the content of scientific theory itself offers no moral foundation for the personal conduct of life.'

Question: 'It is a remarkable fact that although in the Catholic Church and in the Protestant churches, especially of the English-speaking countries, there has always been more or less of a bitter opposition to science, in the highly organized Jewish religion there never has been a spirit of antagonism between religious teaching and scientific investigation.'

Einstein: 'It is quite easy to understand how it has happened that in the history of the Jewish religion there has been no opposition to science. For the Jewish religion is, more than anything else, a way of sublimating everyday existence, and it entails no narrow discipline in doctrinal matters affecting one's personal views of life. As a matter of fact, it demands no act of faith – in the popular sense of the term – on the part of its members. And for that reason there never has been a conflict between our religious outlook and the world outlook of science.'

In some of his writings[27] Einstein has coined the term 'cosmic religion' for his own kind of religious feelings.

> It is very difficult to elucidate this feeling to anyone who is entirely without it, especially as there is no anthropomorphic conception of God corresponding to it.
>
> The religious geniuses of all ages have been distinguished by this kind of religious feeling, which knows no dogma and no God conceived in man's image; so that there can be no church whose central teachings

> are based on it. Hence it is precisely among the heretics of every age that we find men who were filled with this highest kind of religious feeling and were in many cases regarded by their contemporaries as atheists, sometimes also as saints. Looked at in this light, men like Democritus, Francis of Assisi, and Spinoza are closely akin to one another.
>
> How can cosmic religious feeling be communicated from one person to another, if it can give rise to no definite notion of a God and no theology? In my view, it is the most important function of art and science to awaken this feeling and keep it alive in those who are receptive to it.
>
> We thus arrive at a conception of the relation of science to religion very different from the usual one. When one views the matter historically, one is inclined to look upon science and religion as irreconcilable antagonists, and for a very obvious reason. The man who is thoroughly convinced of the universal operation of the law of causation cannot for a moment entertain the idea of a being who interferes in the course of events . . . He has no use for the religion of fear and equally little for social or moral religion . . .
>
> It is therefore easy to see why the churches have always fought science and persecuted its devotees. On the other hand, I maintain that the cosmic religious feeling is the strongest and noblest motive for scientific research . . . A contemporary has said, not unjustly, that in this materialistic age of ours the serious scientific workers are the only profoundly religious people.

We have now arrived at Einstein's ultimate religious opinion: science and religion are not, cannot be, in conflict with each other. Rather, they need each other.

That was the main theme of his address before the Princeton Theological Seminary in May 1939, from which I quote:

> During the last century, and part of the one before, it was widely held that there was an unreconcilable conflict between knowledge and belief. The opinion prevailed among advanced minds that it was time that belief should be replaced increasingly by knowledge; belief that did not itself rest on knowledge was superstition, and as such had to be opposed . . .
>
> The weak point of this conception is, however, this, that those convictions which are necessary and determinant for our conduct and judgments cannot be found solely along this solid scientific way.
>
> For the scientific method can teach us nothing else beyond how facts are related to, and conditioned by, each other. The aspiration toward such objective knowledge belongs to the highest of which man is capable, and you will certainly not suspect me of wishing to belittle the achievements and the heroic efforts of man in this sphere. Yet it is equally clear that knowledge of what *is* does not open the door directly to what *should be*. One can have the clearest and most complete knowledge of what *is*, and yet not be able to deduct from that what

should be the *goal* of our human aspirations. Objective knowledge provides us with powerful instruments for the achievements of certain ends, but the ultimate goal itself and the longing to reach it must come from another source . . . Here we face, therefore, the limits of the purely rational conception of our existence . . .

If one asks whence derives the authority of such fundamental ends, since they cannot be stated and justified merely by reason, one can only answer; they exist in a healthy society as powerful traditions, which act upon the conduct and aspirations and judgments of the individuals; they are there, that is, as something living, without its being necessary to find justification for their existence. They come into being not through demonstration but through revelation, through the medium of powerful personalities. One must not attempt to justify them, but rather to sense their nature simply and clearly.

Later, Einstein added this clarifying remark concerning his deep conviction that religion and science are perfectly reconcilable:

It is the mythical, or rather the symbolic, content of the religious traditions which is likely to come into conflict with science. This occurs whenever this religious stock of ideas contains dogmatically fixed statements on subjects which belong in the domain of science. Thus, it is of vital importance for the preservation of true religion that such conflicts be avoided when they arise from subjects which, in fact, are not really essential for the pursuance of the religious aims.[28]

I conclude this excursion into Einstein's opinions about science and religion with a few lines of what I consider his finest and most comprehensive paper on the subject, the only one, incidentally, which was published in a *scientific* journal.[29] Its origin is a written communication to the Conference on Science, Philosophy, and Religion, held in 1940 at the Jewish Theological Seminary of America, New York. I quote:

Instead of asking what religion is I should prefer to ask what characterizes the aspirations of a person who gives me the impression of being religious: a person who is religiously enlightened appears to me to be one who has, to the best of his ability, liberated himself from the fetters of his selfish desires and is preoccupied with thoughts, feelings, and aspirations to which he clings because of their close superpersonal value. It seems to me that what is important is the force of this superpersonal content and the depth of the conviction concerning its overpowering meaningfulness, regardless of whether any attempt is made to unite this content with a divine Being, for otherwise it would not be possible to count Buddha and Spinoza as religious personalities. Accordingly, a religious person is devout in the sense that he has no doubt of the significance and loftiness of those superpersonal objects and goals which neither require nor are capable of rational foundation . . .

Science can only be created by those who are thoroughly imbued with the aspiration toward truth and understanding. This source of

> feeling, however, springs from the sphere of religion. To this there also belongs the faith in the possibility that the regulations valid for the world of existence are rational, that is, comprehensible to reason. I cannot conceive of a genuine scientist without that profound faith.

And then, Einstein summarized his views in one of the best phrases which in my opinion he has ever written – and with which I heartily concur:

> Science without religion is lame, religion without science is blind.

6. Was Einstein a philosopher?

In an autobiographical sketch, published in 1949, the 67-year-old Einstein has written: 'Science without epistemology is – insofar as it is thinkable at all – primitive and muddled,' warning at the same time of the dangers to the scientist of adhering too strongly to any one epistemological system:

> He [the scientist] must appear to the systematic epistemologist as a type of unscrupulous opportunist: he appears as *realist* insofar as he seeks to describe a world independent of the acts of perception; an *idealist* insofar as he looks upon the concepts and theories as the free inventions of the human spirit (not logically derivable from what is empirically given); as *positivist* insofar as he considers his concepts and theories justified *only* to the extent to which they furnish a logical representation of relations among sensory experiences. He may even appear as a *Platonist* or *Pythagorean* insofar as he considers the viewpoint of logical simplicity as an indispensable and effective tool of his research.[30]

These statements might be taken to indicate that Einstein thought well of philosophy. But look at what he had written five years earlier to the philosopher Benedetto Croce: 'I would not think that philosophy and reason itself will be man's guide in the foreseeable future; however, they will remain the most beautiful sanctuary they have always been for the chosen.'[31] And at what he had said some time in the 1920s: 'Is not all of philosophy as if written in honey? It looks wonderful when one contemplates it, but when one looks again it is all gone. Only mush remains.'[32]

I do not believe one should attempt to deduce from these assorted comments a simple coherent judgement about Einstein's attitude toward philosophy, bearing in mind, in fact, his own words about himself: 'Today's person of 67 is by no means the same as was the one of 50, or 30, or 20. Every reminiscence is coloured by today's being what it is, and therefore by a deceptive point of view.'[33] If I nevertheless open the discussion with my opinion about Einstein and philosophy, I may be excused by the circumstance that he and I often discussed philosophical issues, which has given me some first-hand impressions that perhaps may be of some value.

It is only fair, however, to state that my knowledge of philosophy in the sense of an academic profession all its own is very limited. All that said, here is my dime's worth.

First, what is philosophy? According to the *Encyclopaedia Britannica*, it is a term whose meaning and scope have varied considerably according to the usage of different authors and different ages. The *Oxford English Dictionary* gives nine distinct definitions. The eighth reads: 'A particular system of ideas relating to the general scheme of the universe.' By that definition one has to consider Einstein to be a philosopher *par excellence*. Note also that well into the nineteenth century the term 'natural philosophy' was used for what we now call science. Even today it would be more fitting to call Einstein a natural philosopher than a scientist.

But was Einstein a philosopher in the academic sense of the term? The answer to that question is no less a matter of taste than of fact. I would say that at his best he was not, but would not argue strenuously against the opposite view. It is certain that his interest in philosophy was genuine, and no less certain that his impact on philosophy was profound; yet he did not consider himself a philosopher. It is also certain that his best work was not influenced by any conventional philosophical system. I quite agree with the following judgement – by a philosopher: 'Einstein was a physicist and not a philosopher. But the naïve directness of his questions was philosophical.'[34]

7. Acquaintance with philosophical writings

From the time Einstein was ten until he turned fifteen, Max Talmud, a medical student with little money, came for dinner at the Einsteins every Thursday night. He introduced the young boy to popular books in science and the writings of Immanuel Kant. The two would spend hours discussing science and philosophy. Talmud has recalled:* 'In all these years I never saw him reading any light literature. Nor did I ever see him in the company of other boys of his age.'[35]

I interject Einstein's comments on Kant[36] made during a discussion, in July 1922, in the Société Française de Philosophie:

> I do not think my [relativity] theory accords with the thought of Kant, that is, with what that thought appears to me to be. What appears to me the most important thing in Kant's philosophy is that it speaks of *a priori* concepts for the construction of science. Now there are two opposite points of view: Kant's apriorism, according to which certain concepts pre-exist in our consciousness, and Poincaré's conventionalism. Both agree on this point, that to construct science we need

* In a book he published under the name Talmey. The change of name came after he moved to the United States.

> arbitrary concepts; but as to whether these concepts are given *a priori* or are arbitrary conventions, I am unable to say. . . . In regard to Kant's philosophy I believe that every philosopher has his own Kant.[37]

I return to earlier years. From 1902 to 1909 we find Einstein in Berne, working as a clerk in the Patent Office. In 1903 he, together with two friends, solemnly constituted themselves as founders and sole members of the *Akademie Olympia*. They would dine together, generally have a wonderful time, and discuss philosophy, physics, and literature, from Plato to Charles Dickens. They studied Benedict de Spinoza's ethics, the treatise on human nature by David Hume, the system of logic of John Stuart Mill, and the critique of pure experience by Richard Avenarius. They also read with great care the book *La science et l'hypothése*, by the renowned mathematician and philosopher Henri Poincaré. Maurice Solovine, one of the *Akademie* members, has written: '[This] book profoundly impressed us and kept us breathless for weeks on end.'[38]

Einstein's familiarity with the work of philosophers never stimulated him to write papers that can be called purely philosophical. It did, however, induce him in later years to comment in print on others' philosophical contributions.

From Einstein's introduction to a new German translation[39] of Galileo Galilei's *Dialogue*, we see that he had read Plato: 'The form of dialogue used in his work may be partly due to Plato's shining example.' He wrote an introduction to a new German translation of *De Rerum Natura* by the Roman poet Lucretius.[40] He contributed a commentary to the theory of knowledge in a book[41] on Bertrand Russell. He published two reviews of books on epistemology.[42,43] In one of these he wrote, crisply: 'This book offers an answer to the question what epistemology wants and can do, and especially what it does not want and cannot do.'[42] He also published a critique of a book by Meyerson on the philosophy of relativity in which he commented: 'One must take care not to consider [relativity theory] to be a new mode of thought, distinct from that of the old physics . . . Relativity theory never had such pretensions.'[44] His philosophical interests are also manifest in his introductions to books by Planck[45] and by Frank,[46] and by Viscount Herbert Samuel.[47]

In the obituary following Mach's death, Einstein wrote[48] in laudatory terms of his physics, adding that his intellectual development had not made him into a philosopher. A few years later he expressed himself more emphatically: 'Mach was as good at mechanics as he was wretched at philosophy.'[37]*

* 'Autant Mach fut un bon méchanicien, autant il fut un déplorable philosophe.'

Einstein with the logician Kurt Gödel on Springdale Road, Princeton (early 1950s). (Author's private collection.)

8. Physics and philosophy: relativity theory

The 1922 meeting of the French Philosophical Society, mentioned earlier, was attended by physicists, mathematicians, and the philosophers Henri Bergson, Léon Brunschvicg, Édouard LeRoy, and Émile Myerson. In the course of the discussions of main topics, special and general relativity,* Bergson expressed his admiration for Einstein's work: 'I see [in relativity theory] not only new physics, but also, in certain respects, a new way of thinking.'[49]

Indeed. Relativity of motion had preoccupied philosophers since antiquity, from Aristotle to Newton, Gottfried Leibniz, and Ernst Mach. Kant had raised the issue of the relativity of simultaneity.[34] Special relativity, unveiled in 1905, resolved those problems in ways never seen before. Basic concepts such as the length of measuring rods or the simultaneity of distant events now lost their objective meaning.† In Einstein's own words: 'There is no such thing as simultaneity of distant events.'[51]

Thus special relativity, published in 1905, brought forth first and foremost new ways of thinking in physics itself, but academic philosophers also had to, and did, take notice. That caused a fair amount of confusion, as for example in Bergson's little book on the subject, written as late as 1922.[52]

* Special relativity theory relates the observations of two observers moving with constant, rectilinear velocity relative to each other. General relativity does the same for arbitrary velocities.

† For further details, see Chapter 5, 'A minibriefing on relativity'. For more information see ref. 50.

When a student in Berlin once complained to Einstein about her difficulties in following lectures on the philosophy of relativity, he replied: 'No wonder. The less they know about physics, the more they philosophize. Philosophers used to know all there was to be known of science. In itself philosophy is insubstantial [*die Philosophie ist substanzlos*].'[53] In a report of a lecture on relativity that Einstein gave in London, in June 1921, we read: 'He [E.] deprecated the idea that the new principle was revolutionary . . . There was nothing specially, certainly nothing intentionally, philosophical about it.'[54]

In other circles, spiritists were enchanted with 'the fourth dimension', which, be it noted, was not a new physical concept but rather an important new mathematical tool. Relativity became a term often vulgarized in common language – but not in the witty limerick which expresses in popular terms the relativity principle, according to which the velocity of light *in vacuo* is the largest possible velocity physically attainable:

> There was a young lady from Wight
> Who traveled much faster than light
> She went out one day
> In the opposite way
> And returned on the previous night.

It was in 1915, just about ten years after Einstein had communicated the special theory of relativity, that he wrote to a friend that he was '*zufrieden aber ziemich kaputt*',[55] content but rather worn out. After the most intense and strenuous period of labor in his life, he had found out how to extend the special to the general relativity theory. This, the crowning achievement of his career, was to bring him world fame.*

General relativity has profound implications for our physical as well as philosophical world picture. Space is no longer flat but curved, the forces of gravity fixing the amount of space curvature. In other words, gravity becomes integrated into the geometry of the world in which we live. We can no longer contemplate a theoretical picture of space by itself, only one of space and matter (the seat of gravitational forces) together.†

The mathematical techniques necessary for coping with these problems are, in Einstein's words, 'genuinely complicated.'[57] It should be emphasized that the physical contents of general relativity have still not been fully fathomed, either during Einstein's lifetime or in the years following his death. It is true that since 1915 the understanding of general relativity has vastly improved, our faith in the theory has grown, and no assured limita-

* The origins of Einstein's fame are discussed in 'Einstein and the Press', Chapters 3 and 4.

† For further details, see again Chapter 5, 'A minibriefing on relativity'. For more information see ref. 56.

tions on the validity of the theory have been encountered. Yet no one today would claim to have a full grasp of the rich content of general relativity.

Nor are the answers to the philosophical problems raised by the theory fully in hand today. To the best of my knowledge, philosophical writings on this subject have been quite limited, no doubt largely because the mathematics involved is so tough. Here I shall mention only that general relativity affected acutely Einstein's own outlook on empiricism.*

> I have learned something else from the theory of gravitation: No ever so inclusive collection of empirical facts can ever lead to the setting up of such complicated equations. A theory can be tested by experience, but there is no way from experience to the setting up of a theory. Equations of such complexity as are the equations of the gravitational field can be found only through the discovery of a logically simple mathematical condition which determines the equations completely or (at least) almost completely. Once one has those sufficiently strong formal conditions, one requires only little knowledge of facts for the setting up of a theory.[59]

During his later years, Einstein grew to rely, excessively I think, on logically simple mathematics. I cannot believe that that has done him much good. Here I have in mind his thirty-year-long unsuccessful search for what he termed 'the total field', better known as unified field theory. It is extremely doubtful whether those efforts of his will be of any relevance for the physics of the future.

His idea was this. At the time he completed general relativity, the only known physical forces other than gravity were the electromagnetic forces. Since his theory had unified the structure of space and time with gravity, should this unification not be extended to include electromagnetism as well?

Bravely Einstein set out to find a more embracing world geometry that would achieve this goal – which he never reached. I need not enter here on further details.†

Already, as a young man of 22, Einstein had expressed the sentiments that drove him to this last solitary quest of his: 'It is a wonderful feeling to recognize the unifying features of a complex of phenomena which present themselves as quite unconnected to the direct experience of the senses.'[61] But at the age of 71 he was courageous enough to express doubts about his position, outlined above, regarding the role of empirical facts:

* See also Einstein's essay of 1921 on geometry and experience.[58]

† See ref. 60 for a complete survey of Einstein's attempts at finding a unified field theory.

> The skeptic will say 'It may well be true that this system of equations [describing a unified field theory] is reasonable from a logical standpoint, but this does not prove that it corresponds to nature.' You are right, dear skeptic. Experience alone can decide on truth.[62]

9. Physics and philosophy: quantum theory

Einstein sometimes liked to make a distinction between two kinds of physical theories.[63] Most theories, he said, are constructive; they interpret complex phenomena in terms of relatively simple propositions. Then there are theories of principle. 'Their starting points are not hypothetical constituents but empirically observed general properties of phenomena . . . The theory of relativity is a theory of principle',[63] the class of theories which was dearest to Einstein.

Quantum theory started in 1900 with the pioneering contributions of Max Planck. During the next 25 years it was neither a theory of principle nor a constructive theory. In fact, it was not a theory at all. Rather, in that period it progressed by unprincipled, but tasteful, invention and application of *ad hoc* rules rather than by a systematic investigation of the implications of a set of axioms.

That style of doing physics was not at all to Einstein's liking. Reminiscing about that period, late in life, he has written: 'It was as if the ground had been pulled from under one, with no firm foundations to be seen anywhere upon which one could have built.'[64] Nevertheless Einstein was among the very first to realize that quantum theory signified a development of utmost scientific importance. Great physicist that he was, he made very early some of the most fundamental contributions to quantum physics. In 1905 he posited that under certain circumstances light behaves as if it is built up out of discrete quanta, or photons. In 1906 he laid the foundations for the quantum theory of the solid state.

In 1925 the quantum theory reached its *terra firma*, becoming a theory of principle called quantum mechanics. In his later years Einstein has called this theory 'the most successful physical theory of our period'.[65] Yet he never accepted quantum mechanics as the final answer to the quantum problems. Was it then not enough for him to appreciate how successful quantum mechanics was? No, it was not. As I explain elsewhere in this book,* to Einstein the success of a theory should be acknowledged but did not necessarily constitute sufficient ground for believing in it. As he once wrote to a friend: 'momentary success carries more power of conviction for most people than reflections on principle.'[66]

* See Chapter 4, 'Einstein, Newton, and success'.

Einstein used his perhaps best-known catchy phrase for expressing his objections: 'God does not play dice.' (I have heard him use these words several times.) Occasionally his utterances would be stronger, such as the phrase also quoted earlier: 'It seems hard to look in God's cards. But I cannot for a moment believe that He plays dice and makes use of "telepathic" means (as the current quantum theory alleges He does).'[67] Remarks such as these should not create the impression that he had abandoned active interest in quantum problems, however. Far from it. To the very end of his life he kept pondering about ways of circumventing what he considered to be the weak points of quantum mechanics.

What did Einstein mean by God not playing dice?

I can best illustrate that with the help of a simple example. Consider the collision of two particles. Pre-quantum-mechanical, so-called classical, theory claims that, given the initial positions and velocities of the particles, it is possible to predict their positions and velocities at any later time for any individual collision. Not so, says quantum mechanics. In the words of Einstein's close friend the physicist Max Born:

> One obtains the answer to the question *not* 'what is the state after the collision' but 'how probable is a given effect of the collision' . . . Here the whole problem of determinism arises. From the point of view of quantum mechanics there exists no quantity which in an individual case causally determines the effect of a collision . . . The motions of particles follows probability laws.[68]

In other words, probability – a concept familiar from the odds in the theory of throwing dice! – had become an integral part of the most basic physical description – an entirely novel use of probability in physics.

That was a great conceptual novelty, which demanded not just a new mathematical formulation of physical principles but also a new logic, a new philosophy if you will, that now goes by the name of complementarity.* It implies in particular that a physical phenomenon is defined only if one specifies the experimental arrangement used in making the experimental observation.

'God does not play dice' was Einstein's pungent way of stating that he could never stomach the abandonment of classical determinism and classical causality, nor that our physical knowledge depends on specifying how that knowledge is acquired, how experiments are set up. He has expressed his dissenting view most succinctly as follows: 'Physics is an attempt to grasp reality as it is thought *independently* [my italics] of its being observed.'[69]

* The contents of complementarity are explained in greater detail in Chapter 2, 'Reflections on Bohr and Einstein'.

This standpoint, which he named objective reality,* is evidently irreconcilable with complementarity.

The fifth of the nine *Oxford English Dictionary* definitions of philosophy (already referred to in Section 6) reads: '(Metaphysical philosophy). That department of knowledge or study which deals with ultimate reality, or with the most general causes and principles of things. (Now the most usual sense.)'

Einstein's opinions about quantum mechanics are less those of a physicist than those of a metaphysical philosopher. It could well be that he was thinking about himself when he wrote the following lines about another metaphysical philosopher:

> Although he lived three hundred years before our time, the spiritual situation with which Spinoza had to cope peculiarly resembles our own. The reason for this is that he was utterly convinced of the causal dependence of all phenomena, at a time when the success accompanying the efforts to achieve a knowledge of the causal relationship of natural phenomena was still quite modest.[71]

I have often wondered, why did this man, who contributed so incredibly much to the creation of modern physics, remain so attached to the nineteenth-century view of determinism and causality? – but have never been able to produce a satisfactory answer.

As to the complementarity concept, it is still considered controversial by a small fringe of physicists, though by no means a lunatic one. I believe, however, that their views, like Einstein's, are bound to fade away quietly.

10. Envoi. Einstein's philosophy

My late friend the physicist Richard Feynman has classed scientists among explorers and philosophers among tourists. 'The tourists like to find everything tidy; the explorers take Nature as they find her.'[72] With the exception of his late-in-life views on quantum mechanics, Einstein was an explorer. I have already remarked that none of his papers deals with philosophy in the academic sense. Here and there one does find philosophical reflections in them, however. Out of these I have attempted to distill what may be called Einstein's philosophy.

About science and philosophy. 'I was always interested in philosophy but only in a secondary way. My interest in science was always mainly confined to issues of principle. This serves best to understand my activities and my abstentions.'[73]

* Spelled out in greater detail in his joint paper with Boris Podolsky and Nathan Rosen.[70]

On discovery. 'Discovery is not effected by logical thought, even though the final product is tied to a logical form.'[74]

On the scientific outlook. 'The longing to behold . . . preestablished harmony* . . . [is an] emotional state . . . similar to that of the religious person or the person in love.'[75]

Also on the scientific attitude: 'The scientist finds his reward in what Henri Poincaré calls the joy of comprehension, and not in the possibilities of application to which any discovery of his may lead.'[76]

On simplicity in science. 'In my opinion there is *the* correct path and . . . it is in our power to find it. Our experience up to date justifies us in feeling sure that in nature is actualized the idea of mathematical simplicity.'[77]

On scientific truth. 'Concepts and propositions get "meaning." viz., "content", only through their connection with sense-experiences. The connection of the latter with the former is purely intuitive, not itself of a logical nature. The degree of certainty with which this connection, viz., intuitive combination, can be undertaken, and nothing else, differentiates empty phantasy from scientific "truth". The system of concepts is a creation of man together with the rules of syntax, which constitute the structure of the conceptual systems. Although the conceptual systems are logically entirely arbitrary, they are bound by the aim to permit the most nearly possible certain (intuitive) and complete co-ordination with the totality of sense-experiences; secondly they aim at greatest possible sparsity of their logically independent elements (basic concepts and axioms), i.e., undefined concepts and underived (postulated) propositions.

'A proposition is correct if, within a logical system, it is deduced according to the accepted logical rules. A system has truth-content according to the certainty and completeness of its co-ordination-possibility to the totality of experience. A correct proposition borrows its "truth" from the truth-content of the system to which it belongs.'[78]

On the aims of science. 'Physical theory has two yearnings:

'1. To encompass as much as possible all phenomena and their connections (completeness).

'2. To achieve this on the basis of as few logically independent concepts and arbitrarily assumed relations between these as possible (basic laws, axioms). I will call this the aim of "logical uniformity". I can formulate this second desideratum, crudely but honestly: We do not only wish to know

* An expression of Leibniz's which Einstein particularly liked.

how Nature is (and *how* her processes develop) but also wish, if possible, to arrive at the perhaps utopian and pretentious-seeming goal to know why Nature *is as it is and not otherwise* [E.'s italics]. In this domain lie the highest satisfactions of the scientist.'[79]

On free will. 'Honestly I cannot understand what people mean when they talk about the freedom of the human will. I have a feeling, for instance, that I will something or other; but what relation this has with freedom I cannot understand at all. I feel that I will to light my pipe and I do it; but how can I connect this up with the idea of freedom? What is behind the act of *willing* to light the pipe? Another act of willing? Schopenhauer once said: "*Der Mensch kann was er will; er kann aber nicht wollen was er will*" ("Man can do what he wills but he cannot will what he wills").'[80]

I believe it to be a fitting conclusion to this essay to apply to Einstein what he himself has written about Newton, on the occasion of the bicentennial of his death:

> 'He was placed by fate at a turning-point in the world's intellectual development.'[81]

References

1. A. Einstein, letter to Rosa Winteler, 3 June 1897.

2. I. Rosenthal-Schneider, *Reality and scientific truth*, p. 74, Wayne State University Press, Detroit, 1980.

3. For more on this episode, see *SL*, pp. 113–14.

4. A. Einstein, letter to C. Lanczos, 21 March 1942.

5. H. Dukas, private communication.

6. Maja Einstein, 'Albert Einstein, Beitrag für sein Lebensbild', p. 12, manuscript, copy in Einstein Archive.

7. Ref. 6, pp. 11–12.

8. Ref. 6, p. 13.

9. C. Seelig, *Albert Einstein*, p. 15, Europa Verlag, Zürich, 1960.

10. A. Einstein, in *Albert Einstein, Philosopher–Scientist*, p. 5, Tudor, New York, 1949.

11. A. Einstein, letter to H. Friedmann, 18 March 1929.

12. A. Einstein, letter to M. Grossmann, 2 January 1908.

13. Ref. 9, p. 166.

14. C. Stoll, letter to H. Ernst, 4 March 1909.

15. A. Einstein, letter to J. Laub, summer 1910, undated.

16. Ph. Frank, *Albert Einstein, sein Leben und seine Zeit*, p. 137, Vieweg, Braunschweig, 1979.

17. For more on the Prague period, see *SL*, Chapter 11.

18. *Prager Tageblatt*, 30 July 1912.

19. *Neue Freie Presse*, 5 August 1912.

20. See 'Einstein and the Press'.

21. The letter from the girl (name unknown to me) is dated 19 January 1936. Einstein's answer, dated 24 January 1936, is reproduced in *Weltwoche*, 19 August 1981, p. 37.

22. *Forum*, **84**, 193, 1930.

23. The same article but with improved language is found in A. Einstein, *Ideas and opinions*, p. 8, Crown Publishers, New York, 1982.

24. *NYT*, 9 November 1930. Reproduced in ref. 23, p. 36. The text used here is identical in substance but not in language with the essay 'Cosmic Religion', in A. Einstein, *Cosmic Religion, with other opinions*, p. 43, Covici-Friede, New York, 1931.

25. A. Einstein, *Mein Weltbild*, Querido Publishers, Amsterdam, 1934; reproduced in ref. 23, p. 11.

26. *Forum*, **83**, 373, 1930.

27. Reproduced in ref. 23, p. 41; also in A. Einstein, *Out of my later years*, p. 21, Citadel Press, Secaucus, New Jersey, 1977.

28. *The Christian Register*, June 1948; reproduced in ref. 23, p. 49.

29. A. Einstein, *Nature*, **146**, 605, 1940; reproduced in ref. 23, p. 44.

30. Ref. 10, p. 684.

31. A. Einstein, letter to B. Croce, 7 June 1944.

32. Ref. 2, p. 90.

33. Ref. 10, p. 3.

34. C.F. von Weizäcker, in P. Aichelburg and R. Sexl, *Albert Einstein*, p. 159, Vieweg, Braunschweig, 1979.

35. M. Talmud, *The relativity theory simplified and the formative years of its inventor*, pp. 164–5, Falcon Press, New York, 1932.

36. *Nature*, **112**, 253, 1923.

37. *Bull. Soc. Fran. Philosophie*, **22**, 91, 1922.

38. M. Solovine, ed., *Albert Einstein, lettres à Maurice Solovine*, p. VIII, Gauthier-Villars, Paris, 1956. This book gives the best available description of the *Akademie*, and the best record of what the members read together.

39. A. Einstein, foreword to Galileo's *Dialogue*, transl. S. Drake, University of California Press, Berkeley, 1967.

40. A. Einstein, introduction to *Lukrez, Von der Natur*, transl. H. Diels, Weidmann, Berlin, 1924.

41. A. Einstein, in *The philosophy of Bertrand Russell*, p. 277, ed. P.A. Schilpp, Tudor, New York, 1949.

42. A. Einstein, *Naturw.*, **18**, 536, 1930.

43. A. Einstein, *Deutsche Literaturzeitung*, Heft 1, p. 20, 1924.

44. A. Einstein, *Rev. Phil. de la France*, **105**, 161, 1928.

45. A. Einstein, Prologue to M. Planck, *Where is science going?* Norton, New York, 1932.

46. Ph. Frank, *Relativity, a richer truth*, Beacon Press, Boston, 1950.

47. Viscount Samuel, *Essay in physics*, Blackwell, Oxford, 1951.

48. A. Einstein, *Phys. Zeitschr.*, **17**, 101, 1916.

49. Ref. 37, p. 102.

50. *SL*, chapters 6 and 7.

51. Ref. 10, p. 61.

52. H. Bergson, *Durée et simultanité: à propos de la théorie d'Einstein*, Alcan, Paris, 1922.

53. E. Salaman, *Encounter*, **52**, (4), p. 18, April 1979.

54. *Nature*, **107**, 504, 1921.

55. A. Einstein, letter to M. Besso, 10 December 1915.

56. *SL*, chapters 9–15.

57. Ref. 10, p. 79.

58. A. Einstein, *Geometrie und Erfahrung*, Springer, Berlin, 1921.

59. Ref. 10, p. 89.

60. *SL*, chapter 17.

61. A. Einstein, letter to M. Grossmann, 14 April 1901.

62. A. Einstein, *Scientific American*, April 1950, p. 17.

63. See e.g. *The Times* of London, 28 November 1919.

64. Ref. 10, p. 45.

65. Ref. 10, p. 81.

66. A. Einstein, letter to M. Besso, 24 July 1949.

67. A. Einstein, letter to C. Lanczos, 21 March 1942.

68. M. Born, *Zeitschr. für Phys.*, **37**, 863, 1926.

69. Ref. 10, p. 81.

70. A. Einstein, B. Podolsky, and N. Rosen, *Phys. Rev.*, **47**, 777, 1935.

71. A. Einstein, introduction to R. Kayser, *Spinoza, portrait of a cultural hero*, p. xi, Philosophical Library, New York, 1946.

72. See the chapter 'The explorers and the tourists', in J. Gleick, *Genius: the life and science of Richard Feynman*, Simon and Schuster, New York, 1992.

73. A. Einstein, letter to M. Solovine, 30 October 1924; ref. 38, p. 48.

74. A. Einstein, 'Autobiographische Skizze', in *Helle Zeit, dunkle Zeit*, ed. C. Seelig, Europa Verlag, Zürich, 1956.

75. A. Einstein, *Ansprachen in der Deutschen physikalischen Gesellschaft*, p. 29, Müller, Karlsruhe, 1918.

76. Ref. 45, p. 211.

77. A. Einstein, *On the method of theoretical physics*, Oxford University Press, 1933.

78. Ref. 10, p. 13.

79. A. Einstein, in *Festschrift für Professor A. Stodola*, p. 126, ed. E. Honegger, Orel Füssli Verlag, Zürich, 1929.

80. Ref. 45, p. 201.

81. A. Einstein, *Isaac Newton*, Smithsonian Report, p. 201, 1927, U.S. Government Printing Office, 1928.

Bust of Einstein by Jacob Epstein, 1933. (Hulton Deutsch Collection Limited).

11

Einstein and the press

1

Introduction

As I write these lines, nearly four decades after Einstein's death, his role as a mythical figure in our culture continues undiminished. Articles centering on him still appear in the press. He is the main character in plays and operas, and is pictured in advertisements, some witty, some vulgar. In recent years I have been interviewed or asked for consultations on Einstein programs by television companies from the U.S.A., Great Britain, Germany, France, Holland, and Japan, as well as by various radio programs.

It is evident that Einstein, creator of some of the best science of all time, is himself a creation of the media in so far as he is and remains a public figure. It is therefore of interest, and will be attempted in this essay, to follow Einstein as he was perceived by the press. For this purpose I have used microfilms of the *New York Times*; clippings from the Einstein Archive at Boston University, where I was helped by Robert Schulmann and Ann Lehar; clippings from the Einstein Archive in the Jewish National and University Library in Jerusalem, where Ze'ev Rosenkrantz assisted me; copies of German newspaper items found for me by Dr and Mrs Steinmüller; and various minor sources. My warm thanks to all who lent me a hand.

The following account of Einstein as he appeared in the press is not complete. I have selected only those articles available to me which I found sufficiently interesting. Also, my collection of clippings is rich but of course cannot possibly be exhaustive. Moreover, I have used only items in English, German, and French, since these are the only relevant languages I read with ease.

Einstein was the second scientist whose achievements were widely reported by the world press. The first was Wilhelm Conrad Röntgen, who late in 1895 announced his discovery of what came to be known as X-rays. In 1896 alone more than fifty books and pamphlets and over a thousand newspaper articles on this work saw the light.[1] Remember that regular reporting on science in newspapers began only just about a century ago . . .

Röntgen's name and work is of course well remembered but he has long since ceased to be part of the steady diet served up by science writers in the press. Einstein's continued presence is in fact a unique phenomenon, all the more singular because only a tiny fraction of those who know his name have any inkling of the background for his scientific significance. As a first example of his magical stature as seen by the general public, I quote from an eyewitness report of a meeting in 1921 in a large concert hall in Vienna where Einstein spoke: 'People were in a curious state of excitement in which it no longer matters what one understands but only that one is in the immediate neighborhood of a place where miracles happen.'[2] To Einstein applies *par excellence* the whimsical yet profound definition of a celebrity: a person who is famous for being well-known.

As a sample of Einstein's own response to the publicity centering around him, I should like to draw attention to his delightful short essay 'Interviewers', which is easily accessible.[3]

The beginning of Einstein's mythical role dates from November 1919. I shall follow him in the press beginning with reports dating back to 1902, however, until his death, in April 1955. The years thereafter are treated as well, but only briefly.

Before starting out, I note that the name Einstein has become so widely known that it is being used as a general noun. For example: the Board on Science and Technology for International Development, with offices in Washington, D.C., is developing an EINSTEIN, or Electronic Information Node for Scientific and Technological Exchange, Inquiry, and Networking. This is an electronic mail and bulletin board which they hope to have fully completed by the end of 1993.

2

1902–19

1. Berne

On 5 February 1902, there appeared in a Berne newspaper[4] the following advertisement (in German, of course):

> Thorough private lessons in mathematics and physics for students and schoolboys are offered by Albert Einstein, holder of the teacher diploma of the Federal Institute of Technology [in Zürich], Gerechtigkeitsgasse 32, first floor. Trial hours free of charge.

Einstein was in need of supplementing a small allowance from his family with other funds. In January his daughter Lieserl had been born (out of wedlock). He had moved to Berne in anticipation of obtaining employment at the Patent Office there, a position that did not become firm until the following June.

The advertisement is of some interest because to my knowledge it is the first occasion on which Einstein's name appeared in a newspaper and the only occasion on which he advertised himself in print. At that time there were no other reasons for his name to appear in the press. He had published a few not-too memorable papers and was still virtually unknown in scientific circles.

Then came Einstein's *annus mirabilis*, the year 1905, during which we witness his absolutely stunning creative outburst. The lightquantum hypothesis* was introduced, Brownian motion was interpreted,† a new method for determining the size of molecules was presented (his Ph.D. thesis), and the special theory of relativity§ was set forth.

Any single one of these theoretical discoveries would have sufficed to guarantee Einstein a prominent and lasting position in the history of science. Yet to my knowledge none of these contributions caused even modest mention in the press,** let alone headlines – even though $E=mc^2$ was already there in print for all to see.

Nor did immediate offers for academic positions at once stream in. (The only change in Einstein's status was promotion in 1906 from technical expert third class to second class at the Patent Office.) Einstein's sister Maja has recalled that in fact the appearance of his two relativity papers of 1905 was initially followed by icy silence from the scientific community.[6]

That state of isolation did not last long, however. From 1906 on young physicists began to come to Berne to discuss relativity with its author. From about 1908 onwards Einstein's reputation in scientific circles grew rapidly. That was also the year in which he was admitted as *Privatdozent*†† at the University of Berne.

2. Zürich

Einstein's first faculty position, as associate professor, dates from 1909, when in October he started work at the University of Zürich. The news of his appointment was reported in newspapers in Zürich[7] and Berne,[8] the earliest press reports I have seen about his doings.

Einstein's academic career was now on its way. Also in 1909 he was awarded his first honorary doctor's degree. In 1912 he was for the first time nominated for a Nobel prize, for relativity. (He was awarded that prize in 1922 – for other work.[9])

* According to which, under certain circumstances, light behaves as a stream of particles, called lightquanta or photons.

† The irregular motions of particles (with a typical radius of the order of one thousandth of a millimeter or less) suspended in a liquid.

§ Relativity refers to the connection between the observations of two observers in relative motion. Special relativity deals with the restricted case in which this motion is rectilinear and has constant velocity.

** See ref. 5 for an account of the early reception of relativity theory in various parts of the world.

†† A non-faculty unsalaried position that gave the right to teach. The only remuneration was a small fee paid by each course attendant.

3. Prague

In March 1911 Einstein, his wife Mileva, and their two sons moved to Prague, the ancient capital of the Bohemian kingdom, then part of the Austro-Hungarian empire. At that time Prague had about half a million inhabitants, of whom about one-fifth were German. By imperial decree Einstein had received his first full professor's appointment, at the German Karl-Ferdinand University, in accordance with a proposal dated 16 December 1910, from the Minister of Education to the Emperor. In January 1911 German newspapers in Prague had already reported[10] his imminent arrival. One of those papers, *Bohemia*, wrote further: 'Professor Einstein is a still young scientist but already a leading personality in the school of theoretical physicists. He and Planck* are the founders of the relativity theory in which time is introduced in physical science as a fourth dimension.'[10] A week later one paper wrote that 'our university has secured a young, but in professional circles already very esteemed scholar.'[11] In May the papers announced[12] an evening lecture by Professor Dr Einstein on the relativity principle, to be held before the German *naturwissenschaftlich-medizinische* association.

4. Zürich

Einstein stayed in Prague only from April 1911 until late July 1912.† Several months before his departure, newspapers in Berne[14] and Frankfurt[15] already had announced his appointment to full professor in Zürich, this time at the Elektrotechnische Hochschule (ETH, Federal Institute of Technology). In July a Prague paper expressed regrets: 'Austria now suffers a heavy, even irreplaceable loss in the field of science. Albert Einstein is leaving the German university in Prague . . . It was a glorious act of the Minister of Education when he attracted the scientist to our fatherland.'[16] There follow interesting comparisons with Germany: 'In the German empire Jews cannot be judges or lieutenants, but in the chairs at German universities they enjoy fully equal rights. The number of Israelitic full professors in Germany is at present about twenty-five.' About attracting foreign scientists: 'Who would leave richly endowed institutes or laboratories for the diminutive, poor, incessantly begging Austrian institutions?'[16]

A few days later Einstein said in an interview with a Viennese paper: 'I must stress that I had no reasons for dissatisfaction with Prague. I went to Prague because I had a position in Zürich with low salary . . . Zürich has the advantages to lie at a lake and near mountains, which is of course

* Max Planck had been the first one after Einstein to publish a paper (an important one) on relativity theory.[6]

† For more on the Prague period, see ref. 13.

appealing to a family father . . . In Prague I did not notice any denominational prejudice, as others have surmised.'[17]

In January 1912 the ETH authorities sent their recommendations for a ten-year appointment to the federal Department of the Interior, accompanied by recommendations from famous colleagues, like Marie Curie, who wrote: 'One is entitled to have the highest hopes for him and to see in him one of the first theoreticians of the future.' In the autumn of 1912 Einstein began the next phase of his scientific career. It was to be another brief one, like the one in Prague.

5. Berlin

In the spring of 1913, Max Planck and Walther Nernst, two of Germany's most distinguished senior physicists, came to Zürich for the purpose of sounding out Einstein about his possible interest in moving to Berlin. A combination of positions was held out to him: membership in the Prussian Academy with a special salary to be paid, half by the Prussian government and half by the physics–mathematics section of the Academy from a fund maintained with outside help, a professorship at the University of Berlin with the right but not the obligation to teach, and the directorship of a physics institute to be established. The new institute was to be under the auspices of the Kaiser Wilhelm Gesellschaft, an organization founded in 1911 to support basic research with the aid of funds from private sources. (That institute started its activities in 1917.)

Einstein reacted rapidly and positively to the approach from Berlin. His correspondence from that period makes abundantly clear the principal reasons for his interest in this offer. Neither then nor later was he averse to discussing physics issues with younger colleagues and students; but he had had enough of teaching classes. All he wanted to do was think. The catalogue of Ph.D. theses awarded at the ETH shows that he had acted as *Korreferent** for four theses, all in experimental physics, but had not taken on Ph.D. students in theoretical physics.

Encouraged by Einstein's response, Planck, Nernst, and two other senior colleagues joined in signing a formal *laudatio*, the statement supporting a proposal for membership, which was presented to the academy on 12 June 1913.[18] On 3 July the physics–mathematics section voted on the proposal. The result was twenty-one for, one against.[19] A number of

* The acceptance of an ETH thesis required formal approval by both a principal examiner (*Referent*) and a coexaminer (*Korreferent*). Einstein acted in the latter capacity for the theses of Karl Renger, Hans Renker, Elsa Frenkel, and Auguste Piccard.

arrangements remained to be worked out, but as early as July 1913 Einstein wrote to a friend that he was going to be in Berlin by the spring of 1914.[20]

The press rapidly got wind of Einstein's impending move. In August 1913 a major Berlin paper reported: 'As we hear by cable from Zürich, Dr. Albert Einstein has received a call to Berlin . . . In spite of his youth, Professor Einstein, only 34 years old, is already a scholar of international reputation. He is the founder of a new direction in mathematical [sic] physics which he, at age 24 [actually 26], has given a new basis by his "relativity principle". Professor Einstein, whose articles have aroused admiration also from opponents of his views . . .'[21] In January 1914 the same paper reported that Einstein had been elected and confirmed as full member of the Prussian Academy of Sciences. 'Professor Einstein will move to Berlin on 1 April 1914, in order to devote himself fully to his researches and learned projects, as salaried member of the Academy.'[22] His election to the *Akademie* was approved by Emperor Wilhelm II.

In April 1914 Einstein indeed moved to Berlin. That same month he received an invitation, the first of its kind, to tell something about his area of research for their readers from the editors of the *Vossische Zeitung*, then the leading newspaper in Germany, comparable in stature to the London *Times*. (It was forced to close when the Nazis came to power.[23]) He wrote a nice piece on relativity theory.[24] Three years later he gave an interview to the same paper about an article dealing with the physicist–philosopher Ernst Mach, sent to him for his opinion by his pacifist friend Friedrich Adler, then in jail for having shot and killed Karl von Stürgkh, the Prime Minister of Austria,* who earlier, as Minister of Education, had attracted Einstein to Prague.[25]

By the time Einstein arrived in Berlin, he had already spent several years on attempts at extending his 1905 results concerning the special theory of relativity to what later would be called the general theory of relativity. In the special case he had (as noted above) considered only relative motions of two observers that are rectilinear and do not vary with time. The aim of the general theory was to relate observations of two observers in general, unrestricted, relative motion.

In 1914 he had not yet succeeded in properly formulating the general theory. By that time he already knew, however, that his earlier special theory was like child's play compared to his far more audacious new enterprise. In 1914 he knew that the general theory would profoundly change our picture of the physical world. He knew now that the universe obeys physical laws that could no longer be described in terms of the kind

* Adler was condemned to death but in 1918 received amnesty.

of geometry we all learned at high school; he also knew that, contrary to earlier views, light in general does not propagate along straight paths, and that this 'bending of light' is best observed during a solar eclipse. (More on these subjects in the next section.)

When, on 2 July 1914, Einstein gave his inaugural address as new member of the Akademie in Berlin, he used the opportunity to sketch the status of his new scheme. The *Vossische* had announced this event a week earlier,[26] and reported a synopsis of Einstein's address the day thereafter.[27] It is the first time, to my knowledge, that Einstein's new ideas about relativity appeared in the world press. The term 'general relativity', not yet used here, appeared first in a newspaper article the following November[28] in which another address by Einstein to the Akademie was reported.*

After Einstein had concluded his inaugural speech, Planck rose to note, in courteous terms, that he did not at all believe in Einstein's new research program. He ended by expressing the hope that the expedition planned to observe the solar eclipse of 21 August 1914 would provide information about the bending of light predicted by Einstein. The same hopes are mentioned in a lucid *Vossische* article[30] entitled 'Solar eclipse and relativity theory'. That eclipse did not produce results, however.

Meanwhile, on 1 August, the First World War had broken out. Einstein's war experiences can be summed up most succinctly like this: it was a period during which he suffered considerably from illness, yet continued his creative output.

As the preceding passage shows, in the second decade of this century Einstein had already begun to be treated by the press as a minor celebrity, but only in papers written in German. It was not until November 1919 that the world press began to take notice – and how.

Before coming to that, I should first add a few lines on Einstein the family man.†

In 1903 Einstein had married Mileva Marić, a fellow student from his years of study in Zürich. They had two sons, Hans Albert (b. 1904) and Eduard (b. 1910). The marriage became an increasingly unhappy one, resulting in separation just after the family's arrival in Berlin. Mileva and the two sons returned to Zürich.

While in Berlin, Einstein became increasingly close to his cousin Elsa, who was three years older than him. She had been married and had two

* Earlier the same newspaper had also written about a lecture by Einstein on quantum physics, given to the Deutsche Physikalische Gesellschaft (German Physical Society).[29]

† I treat that subject in detail in Chapter 1, 'In the shadow of Albert Einstein'.

daughters, Ilse (b. 1897) and Margot (b. 1899). She was divorced by the time Einstein settled in Berlin, and had taken her maiden name, Einstein. In 1919 Albert divorced Mileva and, a few months later, married Elsa.

To my surprise, the search for Einstein items in the Berlin press produced an article on Elsa, dating back to 1913.[31] It is not a very flattering report of a public lecture she gave on German poetry, in which she is reported to have spoken in a rather husky voice, of not quite German kind, and not fully developed breeding. Her presentation contained touches of speculation. She is reported to be 'hardly the right interpreter of typically German poets . . . The audience was not sparing with applause and flowers.'

3

*November 1919: Einstein becomes a world figure**

The London *Times* of 7 November 1919 contains an article that begins as follows:

> REVOLUTION IN
> SCIENCE
> **New Theory of the**
> **Universe**
> **Newtonian Ideas**
> **Overthrown**
>
> Yesterday afternoon in the rooms of the Royal Society, at a joint session of the Royal and Astronomical Societies, the results obtained by British observers of the total solar eclipse of May 29 were discussed.
>
> The greatest possible interest had been aroused in scientific circles by the hope that rival theories of a fundamental physical problem would be put to the test, and there was a very large attendance of astronomers and physicists. It was generally accepted that the observations were decisive in the verifying of the prediction of the famous physicist, Einstein, stated by the President of the Royal Society as being the most remarkable scientific event since the discovery of the predicted existence of the planet Neptune.†

I need to indicate the scientific background for this potent statement. After years of struggle, Einstein had finally arrived at his theory of general

* In this section I cannot avoid repeating a few points made in my Einstein biography.[32]

† Discovered in 1846, after various theoretical analyses had predicted its existence and its orbit.

relativity, published in November 1915. This new theory, still considered one of the high points in twentieth-century science, generalizes Newton's classical theory of mechanics, and leads to certain predictions that are at variance with the Newtonian ones. For present purposes it suffices to single out just one of the new results. According to the old theory, light from a far star grazing the edge of the sun is bent by an amount of about 0.85 of a second of arc (a result not due to Newton himself[33]), whereas general relativity predicts 1.7 of a second of arc – twice the Newtonian value. This doubling pitted an old-established giant against a young emerging one. The experimental decision constituted therefore a moment of high drama in the history of thought.

It is a crucial experimental condition for detecting the tiny bending of light that observations be made during a total solar eclipse, when the intense direct light of the sun, overwhelming the minute amount of bent light from the far-away star, is shielded by the moon – whence the reference in the London *Times* to 'results of the total eclipse of May 29 [1919]', results which strongly suggested that Einstein's prediction was in fact the correct one.

Further down the *Times* article one finds a secondary headline:

> SPACE "WARPED".
>
> Until Einstein came along, it had been assumed – most often tacitly – that space is flat, like a book page. General relativity implied, however, that in the neighbourhood of very massive objects, like the sun, space is actually curved, more like the surface of a ball.* It is this space-warp that accounts for the different predictions for the bending of light in the new as compared to the Newton theory.

The next day, 8 November 1919, the London *Times* continued its coverage in an article headlined 'The revolution in science/Einstein versus Newton', in which it was reported that 'The subject was a lively topic of conversation in the House of Commons yesterday, and Sir Joseph Larmor, F.R.S., M.P. for Cambridge University . . . said he had been besieged by inquiries as to whether Newton had been cast down and Cambridge "done in".' On 9 November the subject was discussed for the first time in a Dutch newspaper.[34] The German press picked it up on 23 November.[35]

Most important of all is what happened in the *New York Times*. The precious index volumes of that paper show no mention whatever of Einstein until November 1919. From that time on until his death, not one single year passed without his name appearing in its columns.

It began on 9 November with an article multiply headlined 'Lights all

* The comparisons with a book and a ball refer to two-dimensional 'spaces'. The actual three-dimensional case – the world in which we live – is hard to visualize but fairly easy to formulate mathematically.

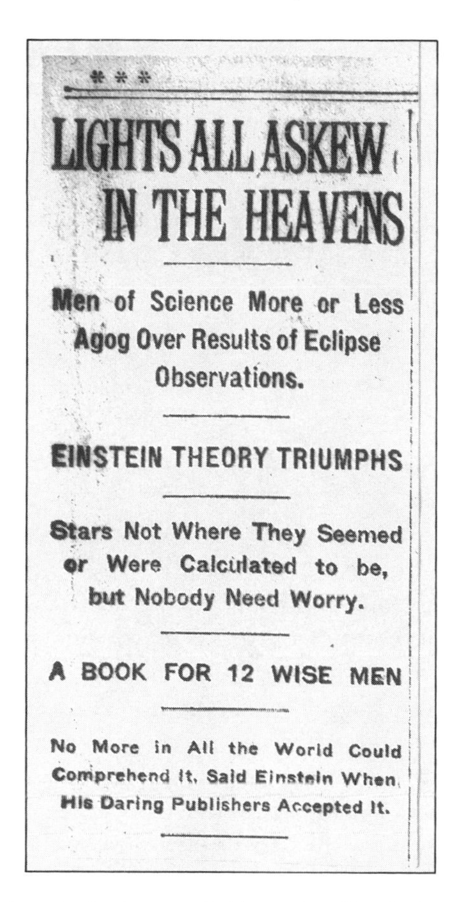

LIGHTS ALL ASKEW IN THE HEAVENS

Men of Science More or Less Agog Over Results of Eclipse Observations.

EINSTEIN THEORY TRIUMPHS

Stars Not Where They Seemed or Were Calculated to be, but Nobody Need Worry.

A BOOK FOR 12 WISE MEN

No More in All the World Could Comprehend It, Said Einstein When His Daring Publishers Accepted It.

'Lights All Askew in the Heavens', November 10, 1919. (©1919 by The New York Times Company. Reprinted by permission.)

askew in the heavens/Men of science more or less agog over results of eclipse observation/Einstein theory triumphs/Stars not where they seem or were calculated to be, but nobody need worry./A book for 12 wise men/No more in all the world could comprehend it, said Einstein, when his daring publishers accepted it.' Again on 11 November: 'This news is distinctly shocking and apprehensions for the safety of confidence even in the multiplication table will arise.' On 16 November: 'These gentlemen may be great astronomers but they are sad logicians. Critical laymen have already objected that scientists who proclaim that space comes to an end* somewhere are under obligation to tell us what lies beyond it.'

On that same 16 November, Charles Poor, professor of celestial mechanics

* In terms of an analogy made earlier, the surface of a sphere is finite, though it has no bounds in or on it.

at Columbia University, said in an interview with the *New York Times*: 'For some years past, the entire world has been in a state of unrest, mental as well as physical. It may well be that the physical aspects of the unrest, the war, the strikes, the Bolshevist uprisings, are in reality the visible objects of some underlying deeper disturbance, worldwide in character . . . This same spirit of unrest has invaded science . . .' There are several more items on relativity that could be quoted from that month's *New York Times*.[36]

4

What caused Einstein's mass appeal?

These various press reports not only make clear that, practically overnight, Einstein had become a heroic figure but they also, it seems to me, contain major clues as to *why* that had happened and why this perception of him has endured to the present day. I make some observations on that issue.

First, Charles Poor's statement places the events just described in the context of their time of occurrence. The drama of Einstein's emergence was enhanced by the exhaustion and chaos following the conclusion, just one year earlier, of the First World War, which had caused millions to die, empires to fall, leaving mankind in a state of uncertainty. Just in these days a new figure appears abruptly, carrying the message of a new order in the universe.

Secondly, the role of language. Even though so very few had and have a real grasp of Einstein's contributions, their contents nevertheless lend themselves very well to expression in everyday language. Take 'space warped'. Everybody thinks he or she knows what 'space' means, everybody knows what a warp is. Yet hardly anyone understands the meaning of a warped space. At the same time, the words are sufficiently familiar to convey to the reader a sense that he has been let in on a highly esoteric piece of information. Likewise with 'stars are not where they seem to be'. Again a phrase that is easily communicated even though the contents carry mystery. I believe that it is precisely this nearness to mystery unaccompanied by understanding which accounts for the universal appeal of the man who created all this novelty. Add to this that stars had for ever been in man's dreams and myths and that irregularities in the skies have of old been considered as omens . . .

My views on the causes of Einstein's mass appeal have been particularly strengthened by his own opinions expressed – only once, as far as I know – in an interview[37] to a Dutch newspaper, given in 1921:

"People slowly accustomed themselves to the idea that the physical states of space itself were the final physical reality."
PROFESSOR ALBERT EINSTEIN

Drawing by Rea Irvin. (©1929, 1957. The New Yorker Magazine, Inc.)

> Does it make a silly impression on me, that excitement of crowds, here and yonder, about my theories of which they cannot understand a word? I think it is funny and also interesting to observe. *I am sure that it is the mystery of non-understanding that appeals to them . . . it impresses them, it has the color and the appeal of the mysterious . . . and one becomes enthusiastic and gets excited.* [my italics]

Nor did this insistence on mystery ever wane. For example, in 1928 the *New York Times* wrote: 'It is a rare exposition of relativity that does not

1930 cartoon of Einstein giving a lecture. (Ullstein Bilderdienst.)

find it necessary to warn the reader that here and here and here he had better not try to understand.'[38]

I do not believe that, as is sometimes imagined, Einstein's looks had much to do with his mass appeal. It is true that he was quite photogenic, but only in his later years. Pictures from the 1920s, when it all began, show him as a friendly looking, pot-bellied gent, dressed in a rather bourgeois way – no doubt under the influence of his second wife, who was bourgeoisie personified.* Sweatshirts came only late in life.

This concludes my remarks on the birth of the Einstein legend. All that follows are variations on this theme, including Einstein's pronouncements on issues scientific, philosophical, and political.

* As told to me by several who knew her personally (I never did) and as is also clear from photographs.

5

The early nineteen-twenties

1. Changes in style

In 1919 Einstein was forty years old and just past the midpoint of his life. Until then his mind had been almost uniquely focused on physics. He had now reached the acme of his creative efforts. In the years to follow, physics would continue to be of overriding importance to him, yet from the time of the events just previously recounted we note a distinct spreading in the subjects that he would write about. The unique focus had gone. Up to 1925 he would continue to produce seminal work. Thereafter it was not all over, but his contributions would become distinctly less influential.

There are two main reasons for what one might call this slackening. First, not even an Einstein can maintain the creative pitch he had been working at during the preceding two decades. Secondly, he had to pay the price for being, much against his own desires, in the limelight. Demands on his time and energies would have come from a variety of directions. Even a synopsis of these[39] would go beyond the aims of the present essay. Here I shall attempt only to follow how these changes reflected themselves in the ways Einstein was covered by the press.

2. Once an oracle, always an oracle

To begin, I note that from now on Einstein's opinion was solicited on all kinds of subjects. The first instance was an interview with the Berlin correspondent of the *Daily Mail*, who wanted to know whether Einstein believed that other planets are inhabited. According to a news report, 'Professor Einstein believes that Mars and other planets are inhabited.'[40] I find that a bit hard to believe, but that is what the papers said.

A few more samples follow of pronouncements by Einstein requested by the press, all from the 1920s, all made in Berlin.

1927. 'Professor Einstein does not favor the abolition of the death penalty . . . He could see no reason why society should not rid itself of individuals proved socially harmful. He added that society had no greater right to condemn a person to life imprisonment than it had to sentence him to death.'[41]

1929. 'In an address at Boston on Sunday [7 April] Cardinal O'Connell warned against Dr. Einstein's theory of relativity as "befogged speculations producing universal doubt about God and his creation", and "cloaking the ghastly apparition of atheism."' Asked for comments, Einstein replied that 'Cardinal O'Connell's assertions left him cold and were devoid of interest.'[42]

Also from 1929, an interview by an American, in Berlin. 'Mrs Einstein's attitude toward her distinguished husband is that of a doting parent towards a precocious child . . . His home showed little of his personality.' Mrs Einstein again: 'He is terribly hard to manage.'

What did Einstein consider the best formula for success in life? 'If A is success, I should say the formula is $A = X + Y + Z$, X being work and Y being play.' 'And what is Z?' 'Keeping your mouth shut.'[43]

3. Expository writing

From 1919 on, there arose increased demand on Einstein for explicatory newspaper articles, beginning with a guest article in the London *Times* of 28 November 1919.[44] He had '[accepted] this opportunity with joy and gratefulness . . . after the lamentable breach in the former international relations existing among men of science.' He concluded a lucid account of his recent work with one of those tongue-in-cheek remarks of which he was so fond: 'Today in Germany I am called a German man of science and in England I am represented as a Swiss Jew.* If I come to be regarded as a *bête noire*, the descriptions will be reversed and I shall become a Swiss Jew for the Germans and a German man of science for the English!'

4. Signing manifestos

After 1919, when Einstein had become a celebrity, he was much sought after for lending his name to political and other causes. The first instance in this period where I found him among originators of a political declaration dates from December 1920. It concerned the treatment of imprisoned political commissars who had participated in organizing the short-lived 'Soviet republic' in Hungary, led by Béla Kun (March–August 1919). After its fall, right-wing radicals had taken over the government and had mercilessly executed many leaders of the Kun régime. This led a considerable number of European politicians and intellectuals to issue a joint declaration calling for 'No death sentences for political crimes', which was published in the press.[45] Among those who signed one finds Einstein, Bernard Shaw, H.G. Wells, Romain Rolland, and Freud.[46] As already

* In 1901 Einstein had become a Swiss citizen. By about 1920 the legal politics of his citizenship had become complex.

said, this was the first occasion after 1920 on which Einstein participated in a political manifesto. Many more were to follow.

5. Antisemitic reactions to Einstein

In December 1919 Einstein wrote to a friend: 'Antisemitism is strong here and political reaction is violent.'[47] The causes are well known. Germany needed scapegoats for its military defeat. Feelings were further kindled by the tens of thousands of Jews who had fled to Germany in order to escape worse fates in Poland and Russia.

In that period Einstein himself became a highly visible target for antisemitic utterances, first because in general fame attracts hatred, secondly because of his advocacy of pacifist and supranationalist causes so alien to the strong German political right.[48]

It appears that the first public demonstration of such sentiments occurred on 12 February 1920. During a popular lecture, which Einstein gave at the University of Berlin, disturbances broke out. A newspaper[49] spoke of 'excesses of an antisemitic mob'. In a statement to the press,[50] Einstein observed that there was a certain hostility directed against him which was not explicitly antisemitic, although its tenor could be so interpreted. The incident was but a foretaste of worse to come.

More serious was the personal attack on Einstein by the *Arbeitsgemeinschaft deutscher Naturforscher zur Erhaltung reiner Wissenschaft* (working group of German scientists for the preservation of pure science). On 24 August 1920, this newly formed organization held a meeting in Berlin's largest concert hall for the purpose of criticizing relativity and the alleged tasteless propaganda methods of its creator.* Einstein himself attended. A few days later he published a fairly detailed retort entitled 'My reply' in the daily press,[52] in which he observed: 'I have good reason to believe that other motives than the search for truth underlie this enterprise. (Had I been a German national with or without swastika instead of a Jew with liberal international opinions, then . . .)' Another letter in the press signed by some distinguished colleagues sharply protested about the proceedings.[53]

On the whole, Einstein's own piece is not well written. He sharply rebutted several persons who were not even worth insulting. (Two weeks later Einstein admitted to a friend that his article had been stupid.[54])

The foreign press also took note of the cabal.[55] Rumors that Einstein might like to leave Germany were correctly denied in Berlin papers.†

* This organization later published a collection of anti-Einstein essays.[51]

† A second meeting held on 2 September 1920 by the same organization attracted little public attention.

6

First journey to the United States and Britain

Einstein paid his first visit to the United States, from 3 April to 30 May 1921, in the company of Elsa, his second wife, and of Chaim Weizmann, head of the World Zionist Organization (later Israel's first president), whose first objective was to garner financial aid for the Zionist cause. Einstein was to raise funds for the planned Hebrew University (which was inaugurated in 1925); and, of course, he intended to visit scientific institutions. Jewish legionnaires who had fought in Palestine were planning to march to Hoboken to meet them as they disembarked from the S.S. *Rotterdam*.[56]

I have heard Abba Eban, the Israeli diplomat, tell the following story. When the *Rotterdam* neared port, the pilot boat came alongside, carrying several reporters who, as is customary, interviewed distinguished travelers while the ship was manoeuvered into the harbor. One of these asked Weizmann whether he had had discussions with Einstein during the voyage. Weizmann replied that for hours on end Einstein had tried to explain to him the principles of relativity theory and that now he (W.) had become convinced that he (E.) had really understood them.

On his arrival Einstein 'timidly faced a battery of camera men. In one hand he clutched a shining briar pipe and with the other clung to a precious violin.' After a welcome by thousands of fellow Jews, they drove by car to the Hotel Commodore. 'The sidewalks were lined nearly all the way uptown with thousands who waved.'[57] His own impression: 'Jews, Jews, nothing but Jews. It was the first time in my life that I saw Jews *en masse*.'[37]

Einstein's American journey was a triumphant progression interspersed with lectures. On 8 April he and Weizmann received the freedom of New York City.[58] On the 11th they were honored at the Metropolitan Opera House for their work in the cause of Zionism.[59] On the 15th Einstein gave the first of three lectures on relativity theory at Columbia University.[60] During this trip he gave all his lectures in German. On 18 April he gave the first of four lectures at the City College of New York.

On 25 April, Einstein, together with a delegation from the National Academy of Sciences, called on President Harding at the White House. Harding 'smilingly confessed that he, too, failed to grasp the relativity idea'.[61] According to Elsa: 'The meeting was a pantomime since President Harding and Professor Einstein could not talk together at their meeting. The President speaks neither German nor French, and my husband does not speak English. The brief meeting consisted of friendly handshakes and joint picture taking.'[62] The next day Einstein addressed the Academy, expressing the hope 'that the field of activity of scientific men may be reunited'.[63] That was in reference to the vindictiveness, from both sides, still lingering three years after the conclusion of the First World War.

On 2 May Einstein arrived in Chicago for three lectures. On the ninth he received an honorary degree from Princeton University. President Hibben quoted lines in German, calling him 'the new Columbus of science'.[64]

On 17 May Einstein and Weizmann arrived in Boston, where they were met by a large crowd at the railway station.[65] The next day they were received by President Lowell of Harvard and also by Governor Cox.[66] On the 25th the two men arrived in Cleveland, where they had to be protected from onrushing crowds at the railway station.[67]

On 30 May the Einsteins sailed for England. In a statement on their departure, Einstein said: 'The respect and admiration that I have always felt for American scientists have been greatly increased as the result of my personal contact with them.'[68]

On 8 June Einstein arrived in Liverpool. He then proceeded to Manchester, where he lectured and received yet another honorary degree. On the 9th he met the Prime Minister, Lloyd George, and the Archbishop of Canterbury.[69]

On 13 June Einstein spoke before a large audience at King's College, London. He was introduced by Lord Haldane, in whose house he stayed. Haldane noted that he had been 'touched to observe that Einstein had left his house [that morning] to gaze on the tomb of Newton in Westminster Abbey', and he characterized Einstein in these words:

> A man distinguished by his desire, if possible, to efface himself and yet impelled by the unmistakeable power of genius which would not allow the individual of whom it had taken possession to rest for one moment.[70]

Shortly after Einstein's return to Berlin, the *New York Times* reported on a small gathering at which, among others, President Ebert was present and during which Einstein reported that the United States was violently anti-German but that indications were that a change of heart was taking place. In England, he said, statesmen and scholars had in mind to bring about renewed friendly relations with Germany.[71]

Some days later another article appeared[72] with the multiple headlines: 'Einstein declares women rule here/Scientist says he found American men toy dogs of the other sex/People colossally bored'. This a second-hand account of an interview on 4 July with a reporter from the Dutch paper *Nieuwe Rotterdamsche Courant*, in which Einstein had been quoted as saying: 'The excessive enthusiasm for me in America appears to be typically American. And if I grasp it correctly the reason is that people in America are so colossally bored, very much more than is the case with us . . . The people are therefore glad when something is given them . . . over which they can enthuse. And that they do, then, with monstrous intensity . . . I found Princeton fine. A pipe yet unsmoked. Young and fresh . . .' Not surprisingly the *New York Times* shortly afterward published a mildly irritated editorial.[73]

Meanwhile, however, Einstein had given a further interview with a Berlin paper,[74] in which he stated: 'I was horrified when I read that newspaper . . . What fills me most when I think back on America is a feeling of gratitude for the warm and cordial reception I received from colleagues, authorities, and private persons.'

As to the German press, many papers recorded Einstein's travels with pride, but there were also nasty comments: 'The whole Einstein trip was a giant bluff . . . He who had done nothing to ease human suffering for the pressing problems of human life was received as a royal highness in the world of the mind . . . important men of science attack the theory of relativity as erroneous.'[75]

7

A trip to France

Einstein's role in Berlin society was marked by intensely ambivalent reactions to him. On the one hand he was claimed by the establishment as one of their most prominent representatives. On the other, he was vilified as the pushy Jew, as events described above illustrate. Both these tendencies were particularly in evidence when in March–April 1922 Einstein was in Paris for the first time. It was a time at which Franco–German relations were still severely strained. The purpose of his visit, arranged by the distinguished French physicist Paul Langevin, Professor at the Collège de France, was to discuss his work with colleagues in physics, mathematics, and philosophy.

Before Einstein's arrival French papers had already carried several items announcing his visit. On the morning after he had arrived in Paris, Langevin wrote in one of them that 'France . . . receives him in a way worthy of her and him.'[76] The astronomer Nordmann wrote: 'The enthusiastic curiosity with which all thinking persons in France look towards Einstein is immense.'[77] The day before, 'all Paris papers had sent their most ardent reporters'[78] to catch him as he came off the train. They were unsuccessful, however, since Langevin managed to whisk him away. 'Einstein's modesty went so far as to forbid his friends to reveal where in Paris he would be staying.'[78] All through his sojourn, Einstein avoided the press as much as he could and refrained throughout from public statements of a political nature.

The main events of Einstein's visit were the four lectures he gave in the great auditorium of the Collège de France, each one followed by debate. 'When he gives his lectures, an honor guard of professors protects him from unbidden persons. On the day of the first conference, when the crowds were particularly large and some feared for nationalistic demonstrations, the former Prime Minister [and brilliant mathematician] Paul Painlevé stood personally at the entrance door to check the letters of invitation.'[79] Later, Einstein received a letter[79a] telling him that the writer, together with two fellow students, had first climbed over a wall and then up an apple tree, just to hear him speak. 'Einstein spoke in French and did not use a manuscript. When he could not find a word he turned to Langevin, who was sitting next to him, and who in a whisper gave him the desired help.'[80]

The meetings were a great success as described by a Berlin correspondent: 'His friendly looks and his smile won all hearts . . . The audience is a curious mixture of celebrities, fashionable society, and studious youths.'[79]

Einstein was all over town. Skimming newspapers, one finds poems on Einstein and Einstein cartoons. 'All newspapers carried his picture, a whole Einstein literature has grown, science and snobism honor the modest scientist . . . He has become the great fashion, academics, politicians, artists, policemen, cab drivers, and pickpockets know when Einstein lectures. *Tout* Paris knows everything and tells more than it knows about Einstein.'[79]

Not all was harmony, however. Einstein was supposed to attend a meeting of the Académie des Sciences but decided to cancel the visit to that august body because he had been warned that there might be unpleasant incidents. A newspaper reported[81] that on the evening of the planned visit the president hid in the nearby library, that thirty members had planned to leave the meeting on Einstein's arrival, and that one member, a general, had said: 'My conscience forbids me to receive a "*Boche*" as long as Germany has not been admitted to the League of Nations'[81] (which did not happen until 1926). This event led the *New York Times* to editorialize: 'Refusal to hear him is childish, and nothing is to be gained by it except derision and pity. And the French do not like it either.'[82]

Elsewhere a French columnist wrote: 'What irritates and perturbs me is that the Germans would have reasons for thinking that vanity and snobism will overwhelm justified rancor and legitimate anger. What irritates me in the case of Einstein is that he is the right ambassador, only too well chosen, to that snobbish imbecility.'[83]

Einstein's own response to his visit was quite positive. In an interview[84] he stressed that he had made no bones about having come as a representative of German science, that there was no lack of good intentions toward improving scientific relations, but that this would demand relinquishing on both sides certain strongly emotional preconceived notions. That German antagonism had also continued to exist can, for example, be seen from the need Einstein felt to plead for closer cooperation between German and French scientists, which he expressed during a meeting in Berlin, in 1922, convened by German pacifists.[84a]

8

The visit to the Orient

1. The assassination of Rathenau

The high season of political killings in Germany started immediately after the First World War. In 1919, Karl Liebknecht and Rosa Luxemburg, Communist leaders, were executed by government troops. In 1921, Matthias Erzberger, a moderate Catholic politician, who at the end of the First World War had signed the armistice after the generals had backed out, was murdered. On 24 January 1922, the Foreign Minister, Walter Rathenau, a brilliant, cultured Jew, was assassinated on the way to his office.

On that occasion Einstein wrote for a newspaper[85] the first of the many sensitive and eloquent obituary notices that he would compose throughout his later life. 'My feelings for Rathenau were and are joyous admiration and gratitude for the hope and consolation he gave me in Europe's current dark days.' He spoke of the special enjoyment of hearing him chat with friends around the table. Decades later, Einstein once quoted to me one of Rathenau's dicta he liked best (and published): 'When a Jew says that he's going hunting to amuse himself, he lies.'[86]

Einstein further wrote in his obituary: 'It is no art to be idealist when one lives in cloud-cuckoo land; he, however, was an idealist even though he lived on earth and knew its smells as few others . . . I had not anticipated that hate, delusion and ingratitude could go that far.'[85] At about the same time, Einstein also correctly diagnosed one of the main causes for the unrest in Germany: 'Everyone here knows that the financial obligations [imposed by the Treaty of Versailles] laid upon the country cannot be fulfilled at their present figure.'[87]

In the following September the German Association of Scientists and Physicians was to celebrate its first centennial. Einstein had been invited to the meeting to give a lecture on relativity theory. In August newspapers carried the following 'upsetting information': he had declined the invitation since he was going abroad for several months.[88] 'Einstein took this sudden decision when he was informed that his name, too, was on the list of victims who were to be done away with by an organization of murderers who had already killed Rathenau . . . Friends and admirers do everything to prevent or at least postpone his return.'

On 8 October 1922, Einstein and his wife left for a five-month trip abroad. 'After the Rathenau murder, I very much welcomed the opportunity of a long absence from Germany, which took me away from temporarily increased danger.'[89] They returned to Berlin in February 1923. Later that year the failed Beer Hall Putsch in Munich would for the first time bring the name Adolf Hitler to broad attention . . .

One quite different event dating from 1922 and worthy of note. In November the *New York Times* reported: 'The Russian communist party condemned the Einstein theory as being, "reactionary in nature, furnishing support of counter-revolutionary ideas", also being "the product of the bourgeois class in decomposition".'[90]

2. Visit to China

The Einsteins' foreign tour began with short visits to Colombo, Singapore, Hong Kong, and Shanghai. The *Singapore Daily*[91] reported a party in their honor attended by about 200 guests. 'All communities and creeds were represented [including] the leading members of the Jewish community and the bishop of Singapore . . . Discourse with the *savant* in German . . . Mrs. Einstein speaks English very well.' Einstein gave a short address in which he spoke of the 'intellectual ambitions, which is one of the finest traditions of our race. (Hear, hear.)'

On 10 November 1922, the Chinese press reported the Einsteins' arrival in Shanghai that day, noting that the local Jews had planned a welcome ceremony, for which 'the doctor' thanked them but declined, saying he needed a rest.[92] The next day a paper noted that Einstein had originally planned to stay for a few more days, but because of obligations in Japan he would leave that morning from Shanghai's Hei-shan dock. Readers were also informed that Einstein had graduated from Yale University 25 years ago (*sic*) and that he was worshipped by students in Europe, America, and Asia.[93]*

3. Five weeks in Japan

On 11 November at 3 p.m., the Einsteins arrived in Kobe.[93] The best picture of their five-week *séjour* in Japan is found in a report[94] from the

* I am indebted to Professor Ge Ge from Beijing for this information on the Chinese press. According to Professor Ge Ge, Einstein visited Shanghai again from 31 December 1922 until 2 January 1923, on his way back from Japan.

German embassy in Tokyo to the Foreign Office in Berlin. 'The press was full of Einstein stories . . . When he arrived at the station there were such large crowds that the police was unable to cope with the perilous crush . . . Einstein remained modest, friendly, and simple . . . At the chrysanthemum festival the center of attention was neither the Empress nor the Prince Regent, everything turned on Einstein.'*

Apart from participating in such festivities, Einstein also gave lectures while in Japan. The most renowned of these is the one he gave in Kyoto, in which he explained how he had come to the special theory of relativity.†

Einstein was on his way to Japan when a telegram arrived in Berlin informing him that he had received the Nobel Prize.

4. The Nobel Prize and the press

Today announcements of Nobel awards make front page news in the world press. It was not always like that. To find the first communication on Einstein's Nobel Prize in the *New York Times*, turn to page 4, the middle of column 2, of its 10 November 1922 edition to find, in its entirety, the following item:

> Nobel Prize for Einstein. The Nobel committee has awarded the physics prize for 1921‡ to Albert Einstein, identified with the theory of relativity, and that for 1922 to Professor Neils [*sic*] Bohr of Copenhagen.

That was the subdued way in which the world was informed of the recognition of the century's two most renowned physicists. Since Einstein was in the Far East, he could not be present in Stockholm for the year's Nobel ceremonies. Bohr did go. On that occasion the two men (who had first met in Berlin in 1920) exchanged respectful and affectionate letters.**

In the 1919 decree of divorce of Albert and Mileva (which I have seen), it had been stipulated that he would give her the Nobel prize money when the prize came. In 1923 the entire award sum ($32,000 in that year's money) was indeed transmitted to her.††

* Japanese press reports are cited in ref. 95.

† The lecture was given in German. One of those in attendance translated and published it in Japanese,[96] which in turn was translated into English.[97]

‡ The award of that prize had been deferred for one year.

** For a detailed account of Einstein's prize, see ref. 98; for Bohr, see ref. 99.

†† Helen Dukas, private communication.

9

The visit to the Holy Land

1. Twelve days in Palestine

On the way back from Japan, the Einsteins arrived in Palestine on 2 February 1923. They were received at the Lemel School in Jerusalem by the Palestine Zionist Executive. 'The road leading to the school was lined on either side by schoolchildren bearing their banners . . . After Professor Einstein had entered the school, there was no holding back the crowd which had assembled outside. The outer gates were stormed, and the crowd burst into the courtyard and tried to force the inner gates, which were stoutly held by three or four stalwarts.'[100] He was welcomed as a man 'who was holding high the banner of the Jewish Renaissance.'[100]

In his reply Einstein said: 'I consider this the greatest day of my life . . . This is a great age, the age of the liberation of the Jewish soul . . . I have recognized that what has been done in our land is a lasting thing.'[100] On 8 February he was named the first honorary citizen of the new city of Tel Aviv.

Einstein gave lectures in a hall on Mount Scopus, the first science lectures to be given in the provisional rooms of the as yet formally to be opened Hebrew University. In introducing him, the President of the Zionist Executive said: 'Mount the platform which has been awaiting you for two thousand years.'[100] After a few opening sentences in Hebrew (a language he did not master[101]), Einstein went on to speak in French on relativity. After his talk, he was thanked by the British High Commissioner for Palestine. 'As [Einstein] stood on the site of the Hebrew University one felt how deep was his regard for the country of his forefathers.'[100]

Einstein's involvement with the Hebrew University was of short duration. In 1924 he edited the first collection of scientific papers of its physics department. In 1925 he accepted membership on the University's Board of Governors. In 1927 the *New York Times* announced, however, that 'Dr. Judah Magnes [the Rector] is reported as giving out a statement telling of the conflict between Professor Einstein and the Board of Directors of the University. The statement further said that the professor is no more a member of the University's presidium.'[102] Einstein was in fact sharply critical of Magnes's leadership.

Einstein also visited the *Technion* (Institute of Technology) in Hadar Hacarmel, Haifa, where he planted two palm trees in a courtyard. These are still flourishing there. Upon his return home, he accepted chairmanship of the newly formed German Committee for the *Technion*, thus becoming the first head of a Technion Friends group anywhere in the world.[103]

2. Einstein's first public comments on Zionism

Einstein was born but not raised as a Jew.[104] As he grew up, he could not fail to be aware of being a Jew, however, if only because of the antisemitism that was pervasive in those years in western culture, if not in virulent, then in insidious, forms. In fact, as a young man of twenty-nine, he once wrote to a friend concerning a personal visit he was thinking of making in support of a position in Switzerland: 'Wouldn't I probably make a bad impression, no Swiss-German, Semitic looks, etc.?'[105]

A prime example of the spirit of those times is found in the final report of the University of Zürich faculty (1909) regarding the appointment of Einstein. That statement followed a strongly favorable declaration to the faculty by a Professor Kleiner, one of its members:

> These expressions of our colleague Kleiner, based on several years of personal contact, were all the more valuable for the committee as well as for the faculty as a whole since Herr Dr Einstein is an Israelite and since precisely to the Israelites among scholars are ascribed (in numerous cases not entirely without cause) all kinds of unpleasant peculiarities of character, such as intrusiveness, impudence, and a shopkeeper's mentality* in the perception of their academic position. It should be said, however, that also among the Israelites there exist men who do not exhibit a trace of these disagreeable qualities and that it is not proper, therefore, to disqualify a man only because he happens to be a Jew. Indeed, one occasionally finds people also among non-Jewish scholars who in regard to a commercial perception and utilization of their academic profession develop qualities which are usually considered as specifically 'Jewish'. Therefore neither the committee nor the faculty as a whole considered it compatible with its dignity to adopt anti-Semitism as a matter of policy† and the information which Herr Kollege Kleiner was able to provide about the character of Herr Dr Einstein has completely reassured us.[106]

It is dubious whether Einstein ever saw this report in his lifetime.

In any event, Einstein's active interest in the destiny of the Jews was not aroused until immediately after the end of the First World War, when he received first-hand evidence of the fate of those Jewish refugees from Poland and Russia who literally came knocking at his door for help. He became incensed by the hostile German reactions to the arrival of these people who had recently escaped worse fates. Moreover, he was 'annoyed by the undignified assimilationist cravings and strivings which I have observed in so many of my [Jewish] friends . . . These and similar happenings have awakened in me the Jewish national sentiment.'[107]

Einstein's first contacts with Zionism date from those same post-War

* . . . *Zudringlichkeit, Unverschämtheit, Krämerhaftigkeit* . . .

† . . . *den 'Antisemitismus' als Prinzip auf ihre Fahne zu schreiben* . . .

years. The one person who more than anyone else contributed to his awakening was Kurt Blumenfeld, from 1910 to 1914 Secretary General of the Executive of World Zionist Organizations, which then had its seat in Berlin, and from 1924 to 1933 President of the Union of German Zionists. Ben Gurion has called him the greatest moral revolutionary in the Zionist movement. It was Blumenfeld who convinced Einstein to join Weizmann on the visit to the United States described earlier. Having done so, he wrote to Weizmann, urging him not to make any attempts to prevail on Einstein to join the Zionist organization,[108] which Einstein in fact never did. Blumenfeld's influence on Einstein is perhaps best seen from the moving letter the latter wrote to him less than a month before his (E.'s) death: 'I thank you belatedly for having made me conscious of my Jewish soul.'[109]

Einstein's first comments in the press about Zionism were stimulated by the first-hand experiences resulting from his visit to Palestine. As noted above, while there he used the phrase 'our land'. He never expressed himself in that way with reference to any other country. He used similar language in his earliest communications in the press on Zionism. In March 1925 he wrote a brief essay[110] entitled 'The mission of our university', on the occasion of the opening of the Hebrew University, in which he said:

> The opening of our Hebrew University on Mount Scopus, at Jerusalem, is an event which should not only fill us with just pride, but should also inspire us to serious reflection.
>
> A special task devolves upon the University in the spiritual direction and education of the laboring sections of our people in the land. In Palestine it is not our aim to create another people of city dwellers leading the same life as in the European cities and possessing the European bourgeois standards and conceptions. We aim at creating a people of workers, at creating the Jewish village in the first place, and we desire that the treasures of culture should be accessible to our laboring class, especially since, as we know, Jews, in all circumstances, place education above all things. In this connection it devolves upon the University to create something unique in order to serve the specific needs of the forms of life developed by our people in Palestine.

Also in that year he published another essay[111] on the issue of Zionism in its generality from which I quote the following:

> Nietzsche has said that one of the peculiarities of the Jewish people consists of their knowing how to realize 'the subtle utilization of misfortune' . . . [Jews ought to] regain without ridiculous arrogance the awareness of the human values they represent . . . Zionism can help them . . . to know themselves less poorly and to become brave . . . Zionism is on its way to create in Palestine a center of Jewish spiritual life. For this one should forever be grateful to its leaders. The existence of this *moral homeland* [my italics] will, I hope, successfully give a plus

> of vitality to a people that has not yet deserved to die out . . . I believe I can maintain that Zionism, an apparent nationalistic movement, has in final analysis significant merit for humanity at large.

This emphasis on the moral and spiritual values of Zionism was to remain Einstein's view for the rest of his life. Later on I shall come back at some length to this subject. Here I note only that, on a visit to Vienna, a year later, Einstein appealed to the Austrian Jews to support the build-up in Palestine.[112]

3. The return home via three weeks in Spain

Returning to early 1923, I now follow the Einsteins on the last phase of their trip around the world, their visit to Spain. How much more can be said about each stage of this journey than is done here can be seen from the fact that a very readable book of nearly 400 pages has been devoted exclusively to their Spanish experiences.[113] They arrived in that country on 22 February 1923.[114] They first visited Barcelona, then went on to Madrid, from where they made excursions to Toledo and the Escorial, spent time in the Prado, and were received in audience by the King and the Queen Mother.[115] Also in Madrid Einstein gave three lectures in French before the Royal Academy of Exact Sciences.[116] A report from Madrid to the Foreign Office in Berlin commented that 'The press daily brought column-long articles on his doings . . . photographers brought his picture in ever changing surroundings . . . caricaturists tried their hand at rendering his striking head.'[117] The Spanish press gave very extensive coverage to the visit.[118]

The Einsteins' journey home included a visit to Zaragoza. They left Spain on or about 15 March,[119] bound for Berlin.

10

Travels to South America

In 1925 Einstein went on his last long journey, this one to South America.

On 24 March he arrived in Buenos Aires, where he was to give two lectures at the University.[120] From there he proceeded to Uruguay, where he spent a week in Montevideo, giving three lectures in French. The German embassy in Montevideo reported: 'During his week's stay he was the talk of the town and the theme of newspapers . . . received by the president of the Republic . . . by his simple, engaging demeanor he left an excellent impression.'[121] After leaving Uruguay he spent a week in Rio de Janeiro, again lecturing. The embassy to the Foreign Office: 'The Brasilian press devoted numerous articles to Einstein's presence* . . . Also in Rio he gained personal sympathy by his unpretentious behavior.'[121] During this trip he managed to prepare a scientific survey article which appeared in Spanish.[123]

On 5 June Einstein was back in Germany.[124] Apart from three later trips to the United States, this was the last major voyage in Einstein's life.

11

Political involvements: the German years

1. During the First World War

In the preceding section it was noted that Einstein's celebrity status had brought numerous requests for signing manifestos of a political character; one example, a declaration condemning death sentences for political prisoners, has been given in Section 5, part 4. I turn next to a more detailed systematic account of his political involvements during his years in Germany (1914–33).

Einstein's scientific productivity was not affected by the deep troubles of the First World War. Those years rank in fact among the most creative of his entire career. The period from 1914–16 was also intellectually the most strenuous in his life. It was the time during which he completed his most important contribution to science, the theory of general relativity.

During the war years Einstein wrote, in all, one book and about fifty papers, an outpouring all the more astounding since he was seriously ill in 1917 and physically weakened for several years thereafter.

These intense activities did not banish from Einstein's mind a genuine deep concern with the tragic events unfolding in the world around him. (Recall that he was in Berlin during the war years.) On the contrary, the period 1914–18 marks the public emergence of Einstein the radical pacifist, the man of strong moral convictions, who would never shy away from expressing his opinions publicly whether these were popular or not. If he at all put his world fame to use, it was to create a forum for speaking out on moral and political issues.

As noted in Section 3, Einstein's years of world renown began in 1919, after the end of the First World War. This is one of two reasons why his activities during that war received little press coverage. The other is that his political pronouncements during that period were most unpopular in Germany and not yet known elsewhere. Since, however, they are so

* South American press reports are cited in ref. 122.

interesting, I shall depart in the next few pages from my aim of describing Einstein as seen exclusively through the eyes of the press, and shall rely on other sources.*

The first time ever that Einstein put his name to a political manifesto was in late 1914. The occasion was a response to the *Aufruf an die Kulturwelt* (manifesto to the civilized world), made public on 11 October 1914, a document often referred to as *Es ist nicht wahr* (it is not true), the opening words of its six main points. Its signatories

> 'loudly raise their voice . . . in protest against the lies and calumnies with which our enemies try to besmirch Germany's clean position'. The manifesto goes on to deny charges of wanton violation of Belgian neutrality, of atrocities committed in Belgium, specifically the alleged pillage of the Belgian city of Louvain, and of violations of international law. Our enemies have no right to call themselves the defenders of European civilization, since 'they have allied themselves with Russians and Serbs, and have unleashed Mongols and Negroes against the white race.'

Actually, as a distinguished historian has written:

> The brutal behavior of the German armies in Belgium was undeniable . . . Belgian resistance infuriated them . . . They took hostages and executed them when they found opposition . . . In the belief that sniping had occurred and that Louvain was full of *franc-tireurs*, the Germans shot a large number of citizens and set the town on fire. The famous old library of the university was entirely destroyed.[126]

This declaration is also known as the 'Manifesto of the 93', after the number of its signatories: 17 artists, 15 scientists, 12 theologians, 9 poets, 7 law professors, 7 historians, 7 professors of medicine, 5 authors, 4 philosophers, 4 philologists, 3 composers, 2 politicians, and one theater director. Famous men of science lent their names: Emil Fischer, Felix Klein, Nernst, Ostwald, Planck, Röntgen.

This document caused bitter, widespread, and long-lasting anger in Allied circles. A year after the War, the French Prime Minister Georges Clemenceau discussed it in an address to the French Senate, calling it 'That shameless manifesto of the so-called intellectuals, yes, real intellectuals, which I consider the greatest crime of Germany, a worse crime than all other acts of which we know.'[127]

In that same year Einstein had this to say about these angry reactions: 'I quite understand this resentment; it has for me a touch of the comic

* The best of which is ref. 125, which deals with Einstein's political commentaries from 1914 until his death.

although hardly as much as the Manifesto itself and its doughty defenders. When a group of people suffer from collective delusions, one should try to deprive these people from any influence; but men of stature and vision cannot long be ruled by hate and passion unless they themselves are sick . . .'[128] This response, to call the episode comic, was his typical reaction, throughout his life, to people with an overblown sense of authority. A story that Einstein himself would occasionally tell with some glee shows that this had already begun in his childhood. At school a teacher once said to him that he, the teacher, would be much happier if the boy were not in his class. Einstein replied that he had done nothing wrong. The teacher answered: 'That is true. But you sit there in the back row and smile, and that violates the feeling of respect which a teacher needs from his class.'[129]

I return to October 1914. Within days after publication of the declaration of the 93, a counter-document was produced, entitled 'Manifesto to Europeans'. It was drafted by Georg Friedrich Nicolai, professor of physiology at the University of Berlin and a noted pacifist. Einstein participated in its composition.[130]

This manifesto emphasized the need for a common world culture.

> Yet, those from whom such sentiments might have been expected – primarily scientists and artists – have so far responded, almost to a man, as though they had relinquished any further desire for the continuance of international relations. They have spoken in a hostile spirit, and they have failed to speak out for peace . . .
>
> We are stating publicly our faith in European unity, a faith which we believe is shared by many; we hope that this public affirmation of our faith may contribute to the growth of a powerful movement toward such unity. The first step in this direction would be for all those who truly cherish the culture of Europe to join forces.[131]

This manifesto was circulated among the faculty members at the University of Berlin. A number of them were in sympathy, yet only four dared sign: Nicolai, Einstein, the astronomer Wilhelm Julius Foerster, and Otto Buek.*

It is noteworthy that Foerster, an avowed pacifist, was also one of the signers of *Es ist nicht wahr*! After the War several who had lent their names explained in writing the pressures exerted on them, one mentioning that he had responded to a telegram that had stressed that speed was of the essence and had advised him not to worry about specific formulations.[133]

In November 1914 the *Bund Neues Vaterland* (New Fatherland League) was founded. Its aims were to promote prompt achievement of peace

* About whom I know only that in the 1920s he was a newspaper correspondent in Argentina.[132]

without annexation and to establish a postwar supranational organization that would make future wars impossible. Einstein was listed among the Bund's founders; he spoke occasionally at its meetings. From the beginning the organization suffered from official harassment. In February 1916 it was enjoined from further activity. It continued to exist clandestinely, however, and then reappeared in the open in 1918.[134]

Also from that year dates a letter from the Chief of Staff of the Command Mark Berlin to the police president of the city of Berlin, advising of the need to keep a careful eye on the activities of pacifists. Appended to the letter is a list of names of known pacifists, on which Einstein had the honor to appear.[135]

A final point of note relating to Einstein's politics during the war years is his correspondence with the French author and pacifist Romain Rolland, who at that time lived in Switzerland. In September–October 1914 he had published a series of articles in the *Journal de Genève*, collected in 1915 in a book *Au dessus de la mêlée* (above the battle), in which he called for France and Germany to respect truth and humanity. Rolland wrote in his diary: 'Einstein, in spite of his lack of sympathy for England, still prefers that England win rather than Germany; England would know better how to let the rest of the world live.'[136]

During the War, Einstein's relations to the German establishment were ambiguous. As a confessed pacifist, he was an outsider. He was an insider in respect to his obligations related to the administration of German science. These he fulfilled conscientiously, some of them with pleasure. As a member of the renowned Preussische Akademie der Wissenschaften, he published frequently in its *Proceedings*, faithfully attended the meetings of its physics section as well as the plenary sessions, often served on its committees, and refereed dubious communications submitted to its *Proceedings*.[137] On 5 May 1916 he succeeded Planck as president of the Deutsche Physikalische Gesellschaft. Between then and 31 May 1918, when Sommerfeld took over, he chaired eighteen meetings of this society and addressed it on numerous occasions. On 30 December 1916 he was appointed by imperial decree to the Kuratorium of the Physikalisch Technische Reichsanstalt, a federal institution, and participated in the board's deliberations on the choice of experimental programs.[138] He held this position until he left Germany. In 1917 he began his duties as director of the Kaiser Wilhelm Institut für Physik, largely an administrative position, the initial task of the institute being to administer grants for physics research at various universities.

Creator of some of the century's finest science, active pacifist, busy administrator – Einstein's activities during the war span an astounding range. Throughout those years he showed his ever-characteristic blend of deep concern and profound detachment, probably best illustrated by what he wrote to a Swiss friend:

> I begin to feel comfortable amid the present insane tumult (*wahnsinnige Gegenwartsrummel*), in conscious detachment from all things which preoccupy the crazy community (*die verrückte Allgemeinheit*). Why should one not be able to live contentedly as a member of the service personnel in the lunatic asylum? After all, one respects the lunatics as the ones for whom the building in which one lives exists. Up to a point, one can make one's own choice of institution – though the distinction between them is smaller than one thinks in one's younger years.[139]

2. The Weimar Republic

On 11 November 1918 the armistice was agreed upon between the Western powers and Germany. Two days earlier Kaiser Wilhelm II had abdicated, and a German republic had been proclaimed, commonly known as the Weimar Republic, after the place where, in 1919, its constituent assembly met.

Einstein greeted these developments with enthusiasm, and expressed his views in remarks to the students of the University of Berlin.

> Our common goal is democracy, the rule of the people. It can only be achieved if the individual holds two things sacred: First, willing subordination to the will of the people, even when the majority is at odds with one's own personal desires and judgments . . . Secondly, all true democrats must stand guard lest the old class tyranny of the right be replaced by a new class tyranny of the left.[140]

The year 1919 was a time of great change in Einstein's life. From January until June he lectured at the University of Zürich. In February he divorced his first wife; in June he remarried (Chapter 2, Section 5). I have already related the events in November which propelled him to a major public figure on the world stage (Chapters 3 and 4).

It was also the year in which Einstein's political activities show a marked increase. In April he joined a commission of six German intellectuals who on their own initiative planned to investigate war crimes. Some time later he wrote about their findings: 'The conviction is slowly gaining ground here that deep wrong was committed by the German army and that the bitter hatred against Germany is justified.'[141]

In June Einstein was among the more than a hundred intellectuals who signed an international appeal drawn up by Romain Rolland. It says in part: 'Let us point out the disasters which have resulted from the almost complete abdication of intelligence throughout the world and from its voluntary enslavement to the unchained forces . . . Mind is no one's servitor. It is we who are the servitors of the mind. We have no other master . . . We know nothing of peoples. We know the People, unique and universal.'[142]

During 1920 Einstein made no public political comments that I know of. Otherwise, in that year his first newspaper interview[40] appeared on matters other than his scientific *oeuvre*. Several unpleasant events occurred: antisemitic disturbances during his lectures, mass meetings against relativity (Chapter 5, Section 5), and, last but not least, the Kapp putsch.*

We have now reached the time where from now on we can again follow Einstein almost exclusively via the press.

In 1921 Einstein gave his first interview – to an American correspondent – on political matters, in which he said in part: 'Science is suffering from the terrible effects of the war, but it is humanity that should be given primary consideration . . . There can be no peace, nor can the wounds of war be healed until internationalism in culture, commerce, and industry is restored . . . Do not omit to state that I am a convinced pacifist, that I believe the world has had enough of war.'[143] A few years later he said in another interview: 'My pacifism is an instinctive feeling. My attitude is not the result of an intellectual theory but is caused by a deep antipathy to every kind of cruelty and hatred.'[144]

3. The League of Nations

This organization, established in 1920, never lived up to the noble aspirations of its founders. It fell into discredit particularly by its inability to prevent Hitler's repudiation of the Treaty of Versailles and Italy's conquest of Ethiopia. The League ceased activities during the Second World War. In 1946 it was replaced by the United Nations.

In 1922 Einstein was invited, and promptly accepted, membership in the League's Committee on Intellectual Cooperation (later called the International Committee on Intellectual Cooperation), news that was reported in German papers.[145] Since he was to represent Germany, which, as a nation, was not yet a League member (it joined in 1926), he became exposed to even more German venom than before.

In January 1923 the French reoccupied the German Ruhr area, the purpose being to put pressure on Germany to pay its war-reparations obligations. This act hastened the complete collapse of the Germany cur-

* This was an unsuccessful attempt led by Wolfgang Kapp, aided by General Erich Ludendorff (Chief of Staff during the First World War), and military units, to force out the Weimar government.

rency. Einstein was furious, and wrote to the Committee announcing his resignation, declaring that the League had neither the strength nor the goodwill for the fulfillment of its tasks. 'As a convinced pacifist I feel obliged to sever all relations with the League.'[146]

In 1924 Einstein changed his mind and rejoined. As he wrote to a fellow committee member, he had been influenced more by a passing mood of disillusionment than by clear thinking.[147] In 1927 he expressed his judgement of the League in more detail, in an interview with *Berliner Tageblatt*: 'With regard to the great problem of world peace, it cannot be denied that the League has been a keen disappointment to many and, further, that people everywhere have sensed its failure to act with courage and good will. Yet, I feel that all well-intentioned people should support this first attempt at restoring order in the sphere of international relations.'

In July 1924 Einstein joined for the first time a meeting of the Committee on Intellectual Co-operation, held in Paris and chaired by the French philosopher Henri Bergson, who addressed Einstein with words of welcome.[148] Other physicist members present were Marie Curie, Lorentz, and Robert Millikan, from the California Institute of Technology (Caltech). Einstein reported on his experiences in a German newspaper: 'I was happy to observe an honest desire to be objective . . . I also had the opportunity to discuss the question of Germany's entry into the League.'[149]

In 1926 Einstein wrote an article for the press[150] of a, for him, quite unusual nature. It deals purely with administrative matters, the purpose being to inform the general public of the working of the Committee of which he was a member. It describes the nature of its seven sections: for general affairs, dealing with the question of improved understanding between nations with the help of education; for international relations between universities; for scientific matters; for legal issues; for literature, particularly the question of the translation of significant books; for the arts; and for the press.

Einstein attended meetings of the Committee in July 1924, 1925, 1926, 1927, and 1930. His absence in 1928 and 1929 was caused by his temporary physical collapse, early in 1928, brought on by overexertion. An enlargement of the heart was diagnosed. He had to stay in bed for four months. He fully recuperated but remained weak for almost a year. 'Sometimes . . . he seemed to enjoy the atmosphere of the sickroom, since it permitted him to work undisturbed.'[151]

The 1930 meeting was the last Einstein attended. When he received an invitation for a meeting in July 1932, he replied that he had assumed that his term of office had expired in 1931, adding that he did not believe he was the person to do useful work for the Committee.[152]

Years later he described the Committee on Intellectual Cooperation like this: 'Despite its illustrious membership it was the most ineffectual enterprise with which I have been associated.'[152]

During the years just discussed, Einstein's opinions on world affairs were frequently sought by the foreign press. Two examples of interviews in 1925 with American papers: on the treaties of Locarno:*

> The Locarno pact is proof that responsible government circles in Europe are now convinced of the need for a European organization on a supranational basis. The pact makes it obvious that traditional prejudices and war-born resentments have so weakened the population as to allow governments to dare take such a step:[153]

On the Far East and Russia:

> The people of the Far East should not be deprived of the possibility of a decent standard of living . . . Japan is now like a great kettle without a safety valve. She has not enough land to enable her population to exist and develop. The situation must somehow be remedied if we are to avoid a terrible conflict.
>
> As for Russia, it seems to me that, in her economic conditions, she has made little progress under her present form of government and has little to show of a constructive nature. Industrial production has declined. But as to the future of Russia, it is, as in all things, difficult as well as unwise to prophesy.[154]

This is a fitting place to relate a little more about Einstein's attitude toward the Soviet Union. In 1923 he was one of the people who signed a call[155] for founding the 'Society of German friends of the new Russia', which he himself joined as a member of its central committee. The call said in part: 'For several years an economic and intellectual evolution is in progress in Russia . . . [about which] most of the German people are hardly aware.' The purpose of the Society (disbanded in 1933) was essentially cultural: to help inform Germans about economic, technical, scientific and artistic developments in Russia. Inevitably the Society's activities were monitored by the German police.[156]

Einstein responded positively to the aims of Russia's October 1917 revolution, yet had reservations about the strategy and tactics of the Communist Party. As he put it some years later: 'In Lenin I honor a man who devoted all his strength and sacrificed his person to the realization of social justice. I do not consider his methods to be proper.'[157] And about Lenin as a thinker: 'Outside Russia Lenin and Engels are of course not

* A series of agreements between Belgium, England, France, Germany, and Italy, signed on 16 October 1925 in Locarno, which included guarantees of borders as well as of a demilitarized zone east of the Rhine.

valued as scientific thinkers and no one might be interested to refute them as such. The same might also be the case in Russia, but there one cannot dare say so.'[158]

Einstein never visited Russia but he did have some correspondence with its leaders. In 1936 he wrote[159] to Vyacheslev Molotov, at that time the chairman of the Council of People's Commissars, asking him to facilitate plans by one of his co-workers for a temporary position in the Soviet Union. The little man with the pince-nez must have replied to the good professor, for shortly afterward Einstein wrote again to Molotov to thank him for his help.[160] In 1947 Einstein wrote to Stalin to plead the case of Raoul Wallenberg, a junior Swedish diplomat in Budapest who, with his staff, had managed to bring about 20,000 Hungarians under the direct protection of the Swedish legation, and who, in 1945, had vanished after having fallen in the hands of the Soviet army: 'As an old Jew, I appeal to you to find and send back to his country Raoul Wallenberg . . . [who], risking his own life, worked to rescue thousands of my unhappy Jewish people.'[161] In reply, an underling stated that he had been authorized to say that a search for Wallenberg had been unsuccessful . . .[162]

4. Varia

As we now move into the later 1920s, Einstein's creative genius has passed its peak. His last seminal discovery dates from 1925 and deals with the so-called Bose–Einstein condensation.[163] To be sure, during the thirty years he still had to live, he never ceased to be engaged in research. While some of it was good, none of it can be called great.

Together with this change in character of his scientific *oeuvre*, one notes a change in his writings. This began, very slowly at first, after Einstein had scaled his ultimate heights with his formulation of the general theory of relativity in late 1915. From 1916 to 1920 we find his early eulogies to recently deceased men of science. These portraits show Einstein's keen perceptions of people and thereby contribute to a composite picture of the man himself. He was of course also an obvious candidate for contributing to commemorative volumes on great men of the past such as Kepler, Newton, and Maxwell.[164]

None of these changes diminished in the least the attention the world press devoted to Einstein; if anything, the contrary is true. Events in his personal life were considered worth reporting about. Also, it seems to me as if Einstein appeared to be more willing to pronounce on all kinds of issues in interviews. I give a few samples, all from the late 1920s.

About his personal life. 'When one evening Einstein took a bath before going to bed but did not reappear after an hour, his wife became worried and opened the bathroom door. There he lay in his tub, deep in thought which he interrupted as if coming out of a dream. "Ah well, I thought I was sitting at my desk."'[165]

About the distracted professor. 'Taking the train in Paris for Berlin, Einstein left his baggage behind him.'[166]

About his state of health (1928). 'Einstein is suffering from an enlargement of the heart.'[167]

Next, samples of topics on which Einstein was interviewed, ranging from soup to nuts.

A direct quote. 'Why do people speak of great men in terms of nationality? Great men are simply men and are not to be considered from [that] point of view . . . nor should the environment in which they were brought up be taken into account.'[168]

On the death penalty. 'Professor Einstein does not favor the abolition of the death penalty . . . He could not see why society should not rid itself of individuals proved socially harmful. He added that society had no greater right to condemn a person to life imprisonment than it had to sentence him to death.'[169]

5. Militant pacifism: Thou shalt not bear arms

Beginning in 1928 and ending, temporarily, in 1933, when Hitler came to power, Einstein was an outspoken supporter of individual war resistance. His pacifism had become more radical, no longer confined to cultural issues, as in earlier years. Rather, he now expressed himself in favor of the principles of universal disarmament and of unconditionally refusing to bear arms. As he wrote to a French colleague, he would not dare to preach this creed to a native African tribe, however, 'for the patient would have died long before the cure would have been of any help to him.'[170]

Einstein's early expressions of his now more drastic position are contained in letters and in manifestos he signed.[171] These, however, were not picked up by the press, which, as best I know, began to take note only in 1930. The first mention, a special dispatch under Einstein's own name, contains these lines:

> It is generally recognized that the policy of maintaining large armaments, pursued by all powers, has proven most harmful to humanity. I assert, moreover, that, under existing conditions, no country would run any real risk by adopting unilateral disarmament. If this were not so, countries which are now inadequately armed or not armed at all would be in an extremely dangerous and precarious situation. This, however, is not the case.[172]

A few months later Einstein co-signed a manifesto which made a prescient statement:

> Whole nations are in peril!
>
> Do you know the meaning of a new war which would use the means of destruction science is ceaselessly perfecting?
>
> Do you know that in the future war will no longer be profitable to anyone, since not only arms, munitions and food depots but all important industrial centers would be targets of attack? This would bring about total destruction of industries.[173]

That was just for starters. I turn next to Einstein's statements expressing specifically his opposition to men bearing arms.

In 1930: 'I have made no secret, either privately or publicly, of my sense of outrage over officially enforced military and war service. I regard it as a duty of conscience to fight against such barbarous enslavement of the individual with every means available.'[174]

Also in 1930: 'The real pacifists, those who are not up in the clouds, but who think and count realities, must give up idle words, and fearlessly try to accomplish something of definite value to their cause . . . To the timid ones who fear imprisonment by their governments I say: "You need not fear imprisonment, for if you get only two per cent of the population of the world to declare in times of peace, 'We are not going to fight; we need other methods to settle international disputes', this two per cent will be sufficient – for there are no jails enough in the world to hold them!"'[175]

In an interview in 1931: 'Let me begin with a confession of political faith: that the state is made for man, not man for the state. This is true of science as well. These are age-old formulations, pronounced by those for whom man himself is the highest human value. I should hesitate to restate them if they were not always in danger of being forgotten, particularly in these days of standardization and stereotype. I believe the most important mission of the state is to protect the individual and make it possible for him to develop into a creative personality.

'The state should be our servant; we should not be slaves of the state. The state violates this principle when it compels us to do military service, particularly since the object and effect of such servitude is to kill people of other lands or infringe upon their freedom. We should, indeed, make only such sacrifices for the state as will serve the free development of men . . .

'Nationalism, presently grown to such excessive heights, is in my opinion intimately associated with the institution of compulsory military service or, to use a euphemism, the militia. Any state that demands military service of its citizens is compelled to cultivate in them a spirit of nationalism, in order to lay the psychological foundation for their military usefulness. The state must idolize this instrument of brute force to the students in its schools, exactly as it does with religion.

'The introduction of compulsory military service is, to my mind, the prime cause for the moral decay of the white race and seriously threatens not merely the survival of our civilization but our very existence.'[176]

In an interview in 1932: 'Professor Einstein came back over and over again to the importance of personal refusal to take part in war as the most essential and practical step in the process of abolishing it. "Because in this way so few can make so big an impression. You don't need many men, but they must be determined." This thought of a great crusade against war by personal resistance is his central message. "Nothing is to be expected from the Disarmament Conference – let us hope that when this has been generally realized, men's minds will turn to more radical methods . . ."

'Difficult problems are no doubt involved, but they are outweighed by the great moral impression that must be made by the fact that men can no longer be compelled to become murderers. "That is more horrible than anything."'[177]

Now to several of Einstein's statements on the role of governments and nations with regard to disarmament:

'Governments are far too dependent on the economic beneficiaries of the war machine to expect from them a decisive step toward the abolition of war in the near future. I believe serious progress can be achieved only when men become organized on an international scale and refuse, as a body, to enter military or war service. The peoples of the world must be made to realize that no government is justified in expecting its citizens to engage in activities which traditional morality considers criminal.'[178]

'Why should men and nations live in a world where they must fear for their very survival? The answer is that they pursue a certain course for their own miserable and temporary advantage and are unwilling to subordinate their selfish aims to the well-being and prosperity of society as a whole . . .'[179]

'A country that restricts or even suppresses freedom of opinion and criticism on political matters, spoken or written, is bound to deteriorate. A citizenry that tolerates such restrictions testifies to, and thereby further increases, its political inferiority.'[180]

Earlier Einstein had already expressed himself forcefully on freedom of expression. After having seen, in America,* the pacifist film *All Quiet on the Western Front*, he commented on its ban in Germany: 'The suppression of this film marks a diplomatic defeat for our government in the eyes of the whole world. Its censorship proves that the government has bowed to the voice of the mob in the street and reveals so great a weakness that a reversal of policy must be emphatically demanded.'[181]

It is only natural that Einstein's concerns would lead him to reflect on the

* In Hollywood in fact, where a special screening had been arranged for his benefit.

impact of science on society. In February 1931 he said: 'In times of war, applied science has given men the means to poison and mutilate one another. In times of peace, science has made our lives hurried and uncertain. Instead of liberating us from much of the monotonous work that has to be done, it has enslaved men to machines; men who work long, wearisome hours mostly without joy in their labor and with the continual fear of losing their pitiful income.'[182]

In a message (August 1931) to the International Conference of the Opponents of War, he appealed to the scientists of the world to refuse to cooperate in the research for the creation of new instruments of war.[183]

In September 1931: 'The fruits of man's inventive genius over the past century might well have made life carefree and happy, had our institutions been able to keep up with the advance of technology. As it is, these hard-won achievements are, in the hands of our generation, as dangerous as a gun wielded by a toddler. Possession of our wonderful means of production has generated want and famine, and not freedom.'[184]

In January 1932: 'The basis of European–American civilization has been badly shaken; in the face of the sinister, intangible power that threatens, a sense of perplexity and fear grips all of us. While we are richer than any generation before us, both in consumer goods and in the means of production, a great part of mankind continues to suffer from dire want. Production and consumption decline more and more, and confidence in public institutions has sunk lower than ever before. It seems as if everywhere, the whole economic organism were incurably diseased . . . It is not in intelligence that we lack for the overcoming of evil, but we lack in the responsible devotion of man in the service of the common weal.'[185] As I write these lines, it is as if I hear a voice of today, not one of sixty years ago . . .

In concluding this section on Einstein's pacificism in 1928–32, I note his assessment of the pacifist movement.

First, a message he sent[186] to a meeting of War Resisters International held in the summer of 1931:

> You represent the movement most certain to end war. If you act wisely and courageously, you may become the most effective body of men and women involved in the greatest of all human endeavors. Those who think that the danger of war is past are living in a fool's paradise. We are faced today with a militarism more powerful and more dangerous than that which brought on the World War. That is what the governments have accomplished! . . . You must convince the people to take disarmament into their own hands and to declare that they will have no part in war or in the preparation for war. You must call on the workers of all countries to unite in refusing to become the tools of interests that war upon life.
>
> This is not the time for temporizing. You are either for war or

against war. If you are for war, then go ahead and encourage science, finance, industry, religion and labor to exert all their power for the arms build-up, to make weapons as deadly as possible. If you are against war, then challenge everyone to resist it to the limits of his capacity. I ask everyone who reads these words to make this grave and final decision.

. . . I have authorized the establishment of the Einstein War Resisters' Fund.

And, finally, two contributions to the American press:

> The pacifist movement in general is not dramatic enough in peacetime to attract great numbers of people. However, the fight against military service will have a dramatic impact because it will inevitably create a conflict by providing a direct challenge to our opponents.
>
> If, in time of peace, members of pacifist organizations are not ready to make sacrifices by opposing authorities at the risk of imprisonment, they will surely fail in time of war when only the most steeled and resolute person can be expected to resist'[187]
>
> There are two ways of resisting war – the legal way and the revolutionary way. The legal way involves the offer of alternative service, not as a privilege for a few, but as a right for all. The revolutionary view involves uncompromising resistance, with a view to breaking the power of militarism in time of peace or the resources of the state in time of war . . . both tendencies are valuable . . . certain circumstances justify the one and certain circumstances the other.[188]

12

More varia; 1928–32

1. Unified field theory

On 4 November 1928 *The New York Times* carried a story under the heading 'Einstein on verge of great discovery; resents intrusion', followed on 14 November by an item 'Einstein reticent on new work; will not "count unlaid eggs."' Einstein himself cannot have been the direct source of these rumors since these stories erroneously mentioned that he was preparing a book on a new theory. In actual fact, he was at work on a short paper dealing with a new attempt at a unified field theory, his so-called theory of distant parallelism. On 11 January 1929, he issued a brief statement to the press stating that 'the purpose of this work is to write the laws of the fields of gravitation and electromagnetism under a unified viewpoint' and referred to a six-page paper he had submitted the day before.[189] A newspaper reporter added the following deathless prose to Einstein's statement. 'The length of this work – written at the rate of half a page a year – is considered prodigious when it is considered that the original presentation of his theory of relativity [on 25 November 1915] filled only three pages.'[190] 'Einstein is amazed at stir over theory. Holds 100 journalists at bay for a week,' the papers reported a week later, adding that he did not care for this publicity at all. But Einstein's name was magic, and shortly thereafter he heard from the distinguished British scientist Arthur Stanley Eddington. 'You may be amused to hear that one of our great department stores in London (Selfridges) has posted on its window your paper (the six pages pasted up side by side) so that passers-by can read it all through. Large crowds gather around to read it!'[191] The 'Special Features' section of the Sunday edition of *The New York Times* of 3 February 1929 carried a full-page article by Einstein on the early developments in relativity, ending with remarks on distant parallelism in which his no doubt bewildered readers were told that in this geometry parallelograms do not close. So great was the public clamor that he went into hiding for a while.[192]

During the years we are about to discuss, Einstein's theory of distant parallelism was his major attempt at innovative science. In 1931 he was wise enough to give up this line of thought – it was a flop. His time of superb scientific creativity had in fact come to an end – nothing to be ashamed of after more than twenty years of the highest achievements. Was his ever-increasing number of public statements and press interviews on non-scientific matters the cause of or the result of a lowering of concentration on science? I cannot answer this question with any certainty but am inclined to believe that it was the result. At the same time Einstein's expressions of views on issues more general than scientific may serve to make evident to a wider audience his great abilities in handling language (especially German), which I hold second only to his scientific talents.

2. The fiftieth birthday

Einstein reached the age of 50 on 13 March 1929. In order to evade ceremonies and celebrations in his honor, he left Berlin on the 11th. 'Only his wife and daughter will spend the day with him quietly.'[193]

Among the many letters of congratulation sent to Einstein's home in Berlin was one by Max Talmey, who then lived in New York. Talmey had been the first to introduce Einstein, at the age of ten, to popular books on science and to Kant. In reply to his good wishes, Einstein sent Talmey a poem in German which, in the original language, appeared in full on the pages of *The New York Times*.[194] I reproduce it here as just one example of Einstein as the witty poetaster. (During his life he wrote quite a number of such rhymes.)

Jeder zeiget sich mir heute
Von der allerbesten Seite
Und von nah und fern die Lieben
Haben rührend mir geschrieben
Und mit allem mich beschenkt
Was sich so ein Schlemmer denkt
Was für den bejahrten Mann
Noch in Frage kommen kann
Alles naht mit süssen Tönen
Um den Tag mir zu verschönen
Selbst die Schnorrer ohne Zahl
Widmen mir ihr Madrigal
Drum gehoben fühl ich mich
Wie der stolze Adlerich
Nun der Tag sich macht ein Ende
Mach' ich auch mein Kompliment
Alles habt ihr gut gemacht
*Und die liebe Sonne lacht.**

As a birthday gift the City of Berlin intended to give Einstein a house near the outskirts. A house was found but, when Mrs Einstein came to look at it, it turned out that the owner did not want to leave. A piece of land in the nearby village of Caputh was chosen next. The city government delayed the decision, however, when a representative of a nationalist party objected. Einstein solved the problem by buying the land himself and building a house on the property – after having written the following to the Mayor of Berlin: 'Since the efforts for the planned birthday present clearly take too long compared to the duration of human life, I beg you to bury the affair definitively.' In 1954 Einstein wrote further: 'The whole affair is quite amusing. I had acquired the land, stimulated by the failed efforts of the City of Berlin . . . a comedy of not quite unintentional errors. The land was confiscated by the Nazis. Since then it has been taken over by holy Russia, which at one time made a feeble effort to give it back to me with its gracious compliments. Soon thereafter, holy Russia changed its mind, however, and no more made a peep.'[195]

3. 1930 in Berlin: two addresses, one interview

June headline: '4000 bewildered as Einstein speaks,'[196] in the Kroll Opera in Berlin, addressing a World Power Conference, which he ended by

* Everyone shows me today/From his very best side/And from near and far the dears/Have addressed me touchingly/And have given me all/What a gourmand can imagine/What for the man getting along in years/Can be requested/ Everyone approaches with sweet sounds/So that the day is beautiful for me/ Even beggars countless in number/Dedicate their madrigal to me/ Therefore I feel elevated/Like the proud eagle/Now that the day comes to an end/I too pay my compliment/You have done everything right/And the dear sun smiles.

describing one of his (failed) attempts at a unified field theory.

August. Einstein opens the seventh German Radio Exhibition. 'The radio broadcast has a unique function to fill in bringing nations together. It can be used for strengthening that feeling of mutual friendship which so easily turns into mistrust and enmity.'[197]

September. Interview[198] with an American reporter in his home in Caputh, where Einstein had pictures of Michelangelo and Schopenhauer hanging in his study. 'One feels a great aesthetic pleasure while working . . . The intellectual pleasure is closely related to the aesthetic one . . . As to artistic and scientific creation, I hold with Schopenhauer that their strongest motive is the desire to escape the rawness and monotony of everyday life so as to take refuge in a world crowded with the images of our own creation.'

Einstein also told of an offer by an American journal 'to write an article on any subject for an amount that would make an ordinary man dizzy.' He said that he regarded this proposal as the height of impudence and that he had told friends of his objections to being treated as a professional boxer or a moving picture star, not as a scientist.

4. Einstein on Shaw and vice versa

On the evening of 27 October 1930, Einstein was to speak at the Savoy Hotel in London for a dinner arranged by the joint British Committee for the Promotion and Economic Welfare of Eastern Jewry. George Bernard Shaw was to address the same audience as well. Both talks were to be broadcast worldwide. Upon Einstein's arrival at Victoria Station, a reporter asked him what he thought about Shaw. Einstein replied. 'I am no judge of art. I only know that I have to thank Mr Shaw for many happy hours. His humour has made people think freely about life and its problems.'[199]

At the end of his evening discourse,[200] Einstein said: 'It rejoices me to see before me Bernard Shaw and H.G. Wells, for whose conception of life I have a special feeling of sympathy.' Whereupon Shaw rose and gave a glowing speech in which he heaped praise on Einstein. 'In London great men can be had six for a penny . . . There is, however, a kind of great man who did not create empires but universes, whose hands are not sullied by the blood of a single human being . . . I cannot endorse the faith of whatever church but can subscribe to the faith of the universe.' Thus Shaw in a rather rambling discourse, ending with: 'My lords, ladies and gentlemen, I raise my glass to our greatest contemporary: Professor Einstein.' Whereupon Einstein replied: 'I personally thank you for the unforgettable words which you have addressed to my mythical namesake who has made my life so burdensome.'[201]

Einstein, Lord Rothschild, and George Bernard Shaw at a dinner in the Hotel Savoy, October 27, 1930. (With permission of United Press International.)

On that trip Einstein took in some sightseeing for, according to a newspaper report, he was found 'staring in childish delight at the spectacle of the opening of Parliament, thrilled to find himself in the gallery of that mediocre assemblage, blanching before the social lions of London . . . a man who is happiest in his own garden, pottering about in baggy pants, completely indifferent to social graces, happier facing a problem in physics than the Press cameras.'[202]

5. First trip to the American West

On 2 December 1930 Einstein and his wife boarded the *Belgenland*, on their way to America's West Coast via the Panama Canal. He had been invited to be guest professor at Caltech (the California Institute of Technology) in Pasadena, California. On hearing that there was a radio phone on board, he declared: 'I hope journalists are not going to call me up in the middle of the ocean to ask me how I slept the night before.'[203]

On the way out, the Einsteins spent 10–15 December in New York City, making the *Belgenland* their temporary home there. After their arrival, Einstein permitted a 15-minute interview on board ship. According to one reporter: 'Einstein looked like a man who would pay a high price to escape . . . Mrs Einstein speaks English fluently. "Bank books? Ach! He knows

Einstein with the soprano Maria Jeritza and others at the Metropolitan Opera in New York, December 14, 1930. (Bilderdienst Süddeutscher Verlag.)

nothing about bank books. I must handle the bank books."'[204]

Einstein was used by now to being the focus of much attention, but even for him the following days must have been unusually hectic. He lunched with editors of the *New York Times*, then addressed a mass meeting in Madison Square Garden in celebration of Hanukkah, and attended a City Hall Ceremony at which New York's flamboyant Mayor Jimmie Walker handed Einstein the keys to the City.[205] On 14 December he visited Rabindranath Tagore, whom he had met earlier in Berlin, and attended a concert at the Metropolitan Opera House, where he met Arturo Toscanini and Fritz Kreisler.[206]

On 15 December he addressed the New History Society at the Ritz Carlton Hotel on the subject of militant pacifism. Later that day he continued his trip westward via the Panama Canal. In his final message to the New York press he said: 'I am leaving New York with a sense of new spiritual and mental enrichment.'[207]

6. Einstein and Rockefeller

On 14 December Einstein had also called on John Davison Rockefeller at his home on 10 West 54th Street. From a press report of this event, published some time later:[208]

> Professor Einstein argued that the strict regulations laid down by his [Rockefeller's] educational foundations sometimes stifled the man of genius. 'Red tape', the Professor exclaimed, 'encases the spirit like the bands of a mummy!' Rockefeller, on the other hand, pointed out the necessity for carefully guarding the funds of the foundations from diversion to unworthy ends or individuals who are not meritorious. Standing his ground against the greatest mind in the modern world, he ably defended the system under which the various foundations were conducted.
>
> 'I,' Einstein said, 'put my faith in intuition.'
>
> 'I,' Rockefeller replied, 'put my faith in organization.'
>
> Einstein pleaded for the exceptional man. Rockefeller championed the greatest good of the greatest possible number. Einstein was the autocrat, Rockefeller the democrat.
>
> Each was sincere; each, without convincing the other, persuaded the other man of his sincerity.

7. Einstein and Chaplin

On arriving in San Diego, Einstein spoke over the radio. 'You have a mode of living in which we find the joy of life and the joy of work harmoniously combined. Added to this is the spirit of ambition which pervades your very being and seems to make the day's work like a happy child at play.' In the same newspaper report,[209] it was also noted that Einstein had once walked into the dining room of the *Belgenland* dressed only in pajamas.

From the moment their boat docked in San Diego,

> the reception accorded Einstein by Californians was one part show business, one part hero worship, and one part genuine affection. Groups of children dressed in blue and white middies serenaded him and thrust wreaths of flowers into his hands, two bands struck up tunes . . . Will Rogers, the noted humorist, really described the whole sideshow when he said just after Einstein returned to Berlin in March of 1931: 'The radios, the banquet tables and the weeklies will never be the same. He came here for a rest and seclusion. He ate with everybody, talked with everybody, posed for everybody that had any film left, attended every luncheon, every dinner, every movie opening, every marriage and two-thirds of the divorces. In fact, he made himself such a good fellow that nobody had the nerve to ask what his theory was.'[210]

I should next like to relate Einstein's contacts during his Pasadena period with the film world of nearby Hollywood, even though none of my information on this interlude stems from press reports but rather from my late friend Helen Dukas, Einstein's faithful secretary from 1928 until his death, who accompanied the Einsteins on this trip. My reason for digressing here from my main purposes is simply that I find these anecdotes so telling.

One day the Einsteins and Helen were invited to a film screening. While they were watching in the cinema, the film suddenly stopped in the middle, the lights went back on, and a woman marched down the aisle heading for Einstein. She addressed him as follows: 'My name is Mary Pickford [a very famous film star from that period]. I am sorry to disturb you but I wanted so much to shake your hand.' Einstein made some polite reply, Miss Pickford left, the theater turned dark again. Whereupon Einstein turned to his wife and asked: 'Who is Mary Pickford?'

Another day a representative of a Hollywood studio came to call on the Einsteins to ask if the professor would be interested to meet some of the great stars? Einstein replied courteously that he could not think of anyone, but Mrs Einstein allowed that her husband was a great admirer of the films of Charles Spencer ('Charlie') Chaplin. The visitor replied that he would inform the Chaplins, who shortly afterward telephoned, inviting Einstein and his entourage for dinner on a certain night. Chaplin would send a car to pick them up. So it came about that Einstein, his wife, and Helen joined Chaplin and Paulette Goddard, his lady friend, for dinner. Also present were William Randolph Hearst, the newspaper magnate, and Marion Davies, his long-time mistress.

That evening Chaplin invited the Einsteins and Helen to join him for the world première of his new film *City Lights*. They accepted. On 30 January 1931 Chaplin came in person to pick them up. As they approached the Los Angeles Theater in Los Angeles, Einstein was amazed to see nearby searchlights scanning the sky and large crowds of people who started to shriek when they stopped in front of the cinema. Einstein turned to Chaplin and asked: 'What does all this mean?' Chaplin replied: 'Nothing.'

On leaving California, Einstein told the press: 'To me the U.S. revealed itself as . . . a new world of commanding interest. It is a world of confraternity, of cooperation, just as Europe is one of individualism. There every man finds his sphere in which he can effectively absorb himself.'[211] By late March 1931 the Einsteins were back in Berlin.

8. The fate of the American Negroes*

The February 1932 issue of *The Crisis*, organ of the National Association for the Advancement of Colored People, contains a contribution by Einstein:

> It seems to be a universal fact that minorities, especially when their individuals are recognizable because of physical differences, are treated by the majorities among whom they live as inferiors. The tragic part of

* As they were called at the time.

such a fate, however, lies not only in the automatically realized disadvantages suffered by these minorities in economic and social relations, but also in the fact that those who meet such treatment themselves for the most part acquiesce in this prejudiced estimate because of the suggestive influence of the majority, and come to regard people like themselves as inferior.

This second and more important aspect of the evil can be met through closer union and conscious educational enlightenment among the minority, and so an emancipation of the soul of the minority may be attained. The determined effort of the American Negroes in this direction deserves every recognition and assistance.

9. On physics and physicists

In reply to the question: 'Does the world exist independent of human consciousness?', Einstein answered: 'The world, considered from the physical aspect, does exist independently of human consciousness.'[212]

In October 1931 Einstein said in an address to an audience gathered in the Berlin planetarium that physicists were generally spurred on not by a desire to increase human comforts or to aid technological advancement but simply because they sought a better understanding of the nature of the universe. 'When I was a boy I planned to be an engineer, but I said to myself, "So much has already been invented. Why should I, too, devote myself to that sort of thing?"'[213]

10. Einstein at a press conference: a sample

On 22 May 1932, Einstein left for Geneva, where a number of pacifists were to gather on the occasion of a Disarmament Conference. They held a press conference at which also Einstein spoke:

> I should like to say that, if the implications were not so tragic, the methods used at the Disarmament Conference could only be called absurd. One does not make wars less likely to occur by formulating rules of warfare. One must start with the unqualified determination to settle international disputes by way of arbitration.

One gets a clear picture of the impression Einstein made in those years from a description of press peoples' reactions to his presence in Geneva:[214]

> All the newspapermen left the conference chambers when the news spread that Einstein had arrived in Geneva. Even some of the delegates decided that a look at the great man was worth more than listening to bacteriological and aerial discussions.
>
> He had no official standing at the Peace Conference. He was no delegate of any power. He was not even an accredited newspaperman. And yet no one . . . questioned his right to be there. No one, except a Balkan delegate, disputed his supreme right to be there:

'Who sent for him?' he asked. 'What does he represent? Whom does that Jew represent?'

The Balkan delegate was hushed up. An American correspondent actually hit the fool on the mouth with the paper pad in his hand.

It was curious to see the silver-haired, heavy-set man walking up the broad stairs of the Peace Palace, with hundreds of people following him at a respectful distance. Years of interviewing had not given the newspapermen that familiarity with Einstein which they assume even with crowned heads . . .

They all stopped two or three steps below him, ranging themselves quietly when he turned, smiled, and said that he would talk to them later on. A young reporter almost fainted with excitement when Einstein asked him for his lighter to rekindle the cigar he was smoking.

He walked into the conference room. The technician of the Aerial Committee, reading a paper at that moment, stopped for a second, then continued. That brief second, however, was an acknowledgment, a more marked acknowledgment of the greatness the man radiated than if all had stopped everything they were doing and applauded him. All eyes were turned toward Einstein. Where he was the world was . . .

A little later Einstein appeared in the pressroom. And again the newspapermen stood aside at a respectful distance. They did not crowd him, shout at him or hurl questions at him. It was the first real sensation these people had had since they had come to Geneva . . .

13

In which Einstein finally leaves Europe for good

1. The rise of Nazism

Just before leaving for his trip to California, Einstein said to Philipp Frank: 'I believe that the present unstable state of affairs in Germany will continue to hold for about ten years. Thereafter it might be good to be in America.'[215] It was only one year later, however, that Einstein wrote in his diary: 'Today, I made my decision essentially to give up my Berlin position.'[216] He was once again on board ship at that time, *en route* to his second stay at Caltech.

The reason for his accelerated decision was the rise of the National Socialist German Workers Party, the Nazis, led by Adolf Hitler, who had first come to the world's attention by his failed attempt (the so-called Beer Hall Putsch in Munich which was supported by elements from the army) on 8–9 November 1923 at unseating the Bavarian government. At that time he was considered to be just a disgruntled nationalist agitator.

In the following years his Party had slowly grown until it had twelve seats in the *Reichstag*. In the September 1930 elections they had made a stunning advance, however, increasing their number to 107. By comparison, on that occasion the Communist Party obtained only 70 seats, having stupidly wasted as much energy fighting the socialists as their far more serious opponents. Hitler, on the other hand, knew how to appeal to the resentments and exasperations of the German nation, eventually leading it to ruin in the process.

On 17 June 1932 Einstein co-signed an appeal to the German Socialist and Communist parties, urging them to join forces in order to stave off Germany's 'terrible danger of becoming fascist'.[217] That warning came much too late. New elections on 31 July gave the Nazis 230 seats, the largest voting bloc in the *Reichstag*.

On 30 January 1933 Hitler came to power as Germany's Chancellor. Shortly thereafter, in March, Einstein, once again in Pasadena, announced in an interview: 'I am not going home.'[218]

This brief sketch of political developments in Germany has led me far beyond the period December 1931–March 1932 of Einstein's second stay at Caltech. Let us turn to his important encounters during those months.

2. The second visit to Caltech

From December 1931 until March 1932 the Einsteins were once more in the United States, spending most of their time again at Caltech. In February 1932 Abraham Flexner came to Pasadena to discuss with faculty members at Caltech his project for a new research center, the Institute for Advanced Study in Princeton, New Jersey (hereafter called the Institute). On that occasion he was introduced to Einstein. The two men discussed the Institute plan in general terms. When they met again in Oxford, in the spring of 1932, Flexner asked if Einstein himself might be interested in joining the Institute. Einstein replied that he was enthusiastic about coming. Formal negotiations started at once, during which Einstein requested an annual salary of $3000. 'He asked . . . "Could I live on less?"'[219] The appointment was approved in October 1932;[220] his salary was set at $15,000 per annum.

3. Einstein and Freud

In the summer of 1932 Einstein and Sigmund Freud exchanged letters which have been published in a booklet, *Warum Krieg?* (*Why War?*), that has often been reprinted.[221]

Einstein, writing first, began his letter (30 July 1932) by explaining to Freud that he had chosen him after the League of Nations had invited him (E.) to select anyone he wanted to discuss a problem of his own choice; and that he had picked the question: is there a way to liberate people from the disasters of war? He believed Freud was best suited to 'shed light on this question because of your deep knowledge of human drives'. He went on to state that in his opinion the way to proceed would be the creation of a supranational organization on which all states should call to settle conflicts and to which they should be legally bound by its decisions. Such international security arrangements demand unconditional renunciation of the states' fully autonomous freedom of action. Recent failures of seriously intended efforts – Einstein continued – had been caused by powerful psychological forces: the leaders' craving for power and the opportunities for economic gains, by leaders in industry for example, resulting from victory in war.

Einstein then raised the question: How can these small power groups make the great masses obedient to their wishes? Because, he believed, they control the schools, the press, and often also religious organizations. But that, he wrote, can still not be the complete answer. Indeed, how can these controls generate the masses' fury and willingness for self-sacrifice? The answer can only be, he thought, because in man lives a need to hate and to destroy. 'That seems to be the deepest problem of all.'

This led Einstein to pose to Freud his main question: are there possibilities for guiding the psychic development of man in such a way that he grows more resistant to the 'psychosis of hate and destruction'?

Einstein concluded as follows: 'It would be most useful if you could present the problem of the pacification of the world in the light of your new insights.'

Freud's reply in September, five times as long as Einstein's letter, begins by stating that he agrees with Einstein, by reaffirming his own pacifism, and by apologizing for the length of his letter, necessitated by the need for 'writing of generally known and recognized matters as if they were new'.

Turning to the subject, Freud noted that killing the enemy is a compulsive tendency (*triebhafte Neigung*). Already in the animal kingdom conflicts of interest are decided by force.

Justice, Freud continues, is the power of a permanently maintained community of individuals. Initially justice was brute force and even today it cannot do without force.

Freud goes on to remind Einstein of the basic tenet of his psychological theory: Man lives by two impulses, one, the erotic–sexual, unifying; the other, aggressive, to destroy and to kill. Therefore *efforts to abolish aggression have no prospect of success* (my italics). This leads Freud to pose a remarkable question of his own: Why do we rebel so strongly against war, you and I and many others; why don't we consider it one among many painful necessities of life? The answer must be, according to Freud, that every man has a right to live. He notes, however, that 'as long as there are nations that are prepared for the ruthless annihilation of others, those latter should be prepared for war.'

Freud finally turns to the role of culture, especially its 'strengthening of the intellect which has begun to keep primitive urges in check and to internalize aggressive inclinations . . . Perhaps it is not a utopian hope that the influences of culture and of the justified fear for the consequences of a future war may bring the waging of war to an end in the foreseeable future.'

Freud concluded with apologies 'if my observations have disappointed you.'

I had not yet read these letters to and from Freud in the years I knew Einstein; otherwise I would certainly have asked him what he thought about the man and his work. In that last respect I can, however, add a few comments. In 1928 Einstein twice declined to co-sign the nomination of Freud for a Nobel Prize in physiology or medicine.[222] In 1949 he wrote this about Freud to an acquaintance: 'The old one had . . . a sharp vision; no illusion lulled him asleep except for an often exaggerated faith in his own ideas.'[223] Finally, Einstein's stepdaughter Margot once told me that she, herself very much interested in Freud's writings, once proposed to Albert that they read some of these together. They made a beginning but very soon Einstein lost interest . . .

4. December 1932: Einstein leaves Germany forever

In September 1932 Einstein wrote to the Ministry of Culture in Berlin regarding his forthcoming tenure at the Institute.[224] His original plan was to spend five months of the year in Princeton* and the rest of the time in Berlin. It never worked out that way.

On 10 December 1932 the Einsteins left Bremerhaven on board the steamer *Oakland*, once again bound for California. This stay was planned as still another visit. Yet the Einsteins this time took with them thirty pieces of luggage, a little excessive for a three months' absence. Note that they departed a few months after the spectacular and frightening victory of the Nazis in the Reichstag. Einstein may well have had forebodings of what was about to happen. At any rate, according to Helen Dukas, as they closed their house in Caputh prior to their journey, Einstein turned to his wife and said '*Dreh' dich um. Du siehst's nie wieder*,' – 'Turn around, you will never see it again.'

During the week before he boarded, Einstein gave an address in the Berliner Philharmonie, on relativity theory.[225] It was to be his last public appearance in Germany. On 30 January 1933, while Einstein was in

* The Institute is open the whole year round but formally in session only from late September until Christmas and from late January until early April.

Pasadena, Hitler came to power. As already mentioned on p. 188, in March 1933 Einstein stated publicly that he would not return to Germany. That decision cannot have been made at the moment Einstein heard about Germany's new regime, for on 2 February he sent a letter from Pasadena to the secretary of the Akademie in Berlin concerning his salary arrangement.[226]

5. 28 March–7 October 1933: last European séjour, in Belgium and England

Einstein had to return to Europe since he had obligations there and because arrangements had to be made for the move to Princeton, which now, of course, was to be his only home.

His way back led him as usual to New York where, on 15 March he attended a dinner in his honor, under the auspices of the American Friends of the Hebrew University and the Jewish Telegraph Agency. Nearly a thousand people were there. The report of the event is curious because, first, 'representing the German Ambassador was Dr Otto Krep, German Consul General in this city'. Secondly, when Einstein was asked whether he would remain permanently out of Germany, he replied: 'I have no definite intentions. It all depends on the situation.'[227]

Family and friends had helped the Einsteins find a temporary European abode, the Villa Savoyard in Le Coq sur Mer on the Belgian coast. The S.S. *Belgenland* brought them to Antwerp, where they landed on 28 March. On board ship, Einstein had already written his letter to Berlin resigning from the Akademie.[228]

The Nazis' anti-Einstein campaign had already begun while Einstein was still on the high seas. The *New York Times* of 20 March reported that 'one of the most perfect raids of recent German history was carried out 'when the SA* raided Einstein's Caputh home to search for weapons allegedly hidden there by communists.' According to the *New York Times*, all they found was a bread knife.

According to the German newspapers, Einstein had, also while on the *Belgenland*, prepared a letter to the German embassy in Brussels in which he requested information on how to relinquish his German citizenship, which he had received automatically as a result of his Akademie appointment. On 2 April these papers reported first that, the day before, the political police had blocked the bank account of 'the Jewish professor Einstein';[229] second, a statement released by the press service of the Akademie:

* *Schutzabteilung*, a paramilitary Nazi organization.

> The Prussian Academy of Sciences has indignantly taken note of newspaper reports of Albert Einstein's participation in a smear campaign in America and France. The Akademie has at once demanded an explanation from him. Meanwhile, Einstein has explained his resignation from the Akademie on the grounds that he can no longer serve the Prussian state under the present circumstances. The Prussian Akademie feels Einstein's inflammatory behavior painful, since it considers itself most intimately associated with the Prussian state . . . For this reason it has no reasons for regretting Einstein's departure.[228]

A few days later, on 5 April, Einstein wrote a letter to the Akademie[229] in which he declared that he had never participated in inflammatory activities; that all he had stated to the press was that he did not wish to live in a state where there were no equal rights for all individuals and no freedom of speech; and that he had declared the present situation in Germany to be a state of psychic illness.[230]

So much for Einstein's contacts with the *Herrenvolk*.

While he was in Belgium, offers began to reach Einstein for academic positions in several countries. Chaim Weizmann asked him to come to Jerusalem, which Einstein refused outright because he was highly critical of the Hebrew University's administration. He was approached by Leiden and Oxford, by Madrid and Paris.

The official journal of proceedings in the French parliament reports[231] an address by one of its members, the eminent mathematician Émile Borel, who reminded his colleagues of Einstein's lectures at the Collège de France and the Institut Henri Poincaré; of the 'indescribable ovation' following Einstein's receiving the doctorate *honoris causa* from the University of Paris. 'It is the same unanimity which, I am sure, will be found in the French Parliament (*applaudissement unanime*, unanimous applause).' Whereupon the government proposal to offer Einstein a chair at the Collège de France was unanimously adopted. Not all Frenchmen were equally enthusiastic. One wrote to a newspaper: 'The Collège de France was not created for hospitalizing all those Israelites who, believing to be persecuted, harp on a science inaccessible to other mortals . . . Professor Einstein is a bolshevik and a militant communist.'[232]

On 8 April 1933 a letter from the Ministry of Education was sent to Einstein asking him to make known whether he would accept the offer, 'in conformity with French historical precedent'.[232] Einstein accepted, after a fashion. In May he replied: 'My nomination to the Collège de France has filled me with joy and gratitude. I appreciate this nomination all the more because it expresses a spirit which, as much as one can hope, will save European culture from the grave peril that currently threatens it.'[232]

At about the same time an offer came from the Socialist minister of public affairs in Madrid for the creation of an 'Extraordinary Chair' for Einstein.* Moreover, the Spanish government had agreed to establish an Einstein Institute for relativistic physics in Madrid, to be directed by a refugee scholar named by Einstein himself, who would head the institution in name only. There exists in the Einstein Archive a draft of a letter from April 1933 (undated) by Einstein to the Spanish Minister of Education in which he expresses 'a special joy to have available an opportunity to participate personally in the scientific life of your country'.

Press comments were mixed. Liberal papers attached positive value to Einstein's Jewishness; conservative papers bitterly emphasized the contrast between the treatment of Einstein and that of certain conservative Catholic scientists.

In a letter sent from Le Coq on 17 April 1933 to his colleague Painlevé, Einstein expressed his gratitude for the French offer but declined it:

> I can imagine that the generous measure to offer me a chair at the Collège de France was at least mainly your idea. Therefore I am moved to express my gratitude and joy to you before this matter has officially been brought to a close. The significance of this measure by far surpasses me as a person; rather it is an admission of the French authorities and the French nation to remain true to the old traditions of tolerance and freedom in a time in which these most precious achievements of European culture are seriously threatened in countries previously considered pillars of these supreme principles. I am upset that my time is tied up for a long time to come so that I will not be able to work in Paris for the time being. Until then, the importance of your initiative and my profound gratitude towards you and all those who have translated it into action remain unchanged. With gratitude and long standing affection I am sending you very best wishes.

I have been unable to locate a letter from Einstein formally declining the offer from Spain.

Meanwhile his family had been able to save Einstein's papers in Berlin; these were sent to the Quai d'Orsay by diplomatic pouch. Furniture from his Berlin home was prepared for shipment (and arrived safely in Princeton some time later). Einstein traveled, to Brussels, Zürich, Oxford, Glasgow. His two stepdaughters and also Helen Dukas joined the Einsteins in Le Coq. Their establishment was completed by two guards assigned by the Belgian government who were to watch over their safety. Rumors were rife of planned attempts on Einstein's life.

A reporter from Vienna has described the situation at the Villa Savoyard like this.[234] 'The master of the house remains invisible to visitors, re-

* For the Spanish connection see ref. 233.

porters, photographers . . . Every day he sits bent over his research papers.* He shows neither spitefulness nor bitterness about his personal fate . . . to the Nazis he is stateless person number 14 . . . he is not affected by the Nazi threats to his life which appear not even to interest him.' He gets aroused only when the discussion turns to the political situation. 'I cannot understand the passive response of the whole civilized world to this modern barbarism. Does the world not see that Hitler is aiming for war?'

From another part of the same interview, published[235] in the American press:

> I rely with assurance on one fact which seems to me to imply the swift and inevitable collapse of the nazi dictatorship. It is not the strength or the virtues of the adversaries of nazism on which I count. It is on the stupidity of the nazis themselves . . .
>
> It has been said that the existence of a state of siege permits the worst imbeciles to govern a land. It is not true. Without some intelligence not even a dictator flanked by bayonets can maintain his rule indefinitely. Hitler and his minions lack even that minimum degree of intellectual ability required by a dictatorship under modern conditions . . .
>
> I am a convinced democrat. It is for this reason that I do not go to Russia although I have received some very cordial invitations. My voyage to Moscow would certainly be exploited by the rulers of the Soviets to the profit of their own political aims. Now I am an adversary of Bolshevism just as much as of Fascism. I am against all dictatorships.

Since Einstein foresaw Nazi aggression, what had become of his pacificism? This is what he wrote to a pacifist in Belgium:[236]

> What I shall tell you will greatly surprise you. Until quite recently we in Europe could assume that personal war resistance constituted an effective attack on militarism. Today we face an altogether different situation. In the heart of Europe lies a power, Germany, that is obviously pushing toward war with all available means. This has created such a serious danger to the Latin countries, especially Belgium and France, that they have come to depend completely on their armed forces. As for Belgium, surely so small a country cannot possibly misuse its armed forces; rather, it needs them desperately to protect its very existence. Imagine Belgium occupied by present-day Germany! Things would be far worse than in 1914, and they were bad enough even then. Hence I must tell you candidly: Were I a Belgian, I should not, in the present circumstances, refuse military service; rather, I should enter such service cheerfully in the belief that I would thereby be helping to save European civilization.
>
> This does not mean that I am surrendering the principle for which I have stood heretofore. I have no greater hope than that the time may not be far off when refusal of military service will once again be an effective method of serving the cause of human progress.

* While at Le Coq, Einstein completed two short scientific papers.

Since Le Coq was too close to the German border for his safety, Einstein left the Continent for good on 9 September and went to England. On 3 October, in London, after a few quiet weeks in the country interrupted by calls on Winston Churchill and Lloyd George, he made his first major public appearance since the Nazi takeover. 'Reading quietly in imperfectly spoken English',[237] he spoke at a mass meeting at the Royal Albert Hall (attendance: 10,000) to help raise $5 million for aid to German refugee scientists. The Hall was heavily guarded by Scotland Yard, which had received a tip of a plot to assassinate Einstein. Other speakers included the physicists James Jeans and Ernest (now Baron) Rutherford. Here is part of what Einstein said:

> You have shown that you, and the British people as a whole, have remained faithful to the tradition of tolerance and justice which your country has proudly upheld for centuries.
>
> It is precisely in times of economic distress, such as we experience everywhere today, that we may recognize the effectiveness of the vital moral force of a people. Let us hope that, at some future time, when Europe is politically and economically united, the historian rendering judgment will be able to say that, in our own days, the liberty and honor of this continent were saved by the nations of Western Europe; that they stood fast in bitter times against the forces of hatred and oppression; that they successfully defended that which has brought us every advance in knowledge and invention: the freedom of the individual without which no self-respecting individual finds life worth living.
>
> It cannot be my task to sit in judgment over the conduct of a nation which for many years counted me among its citizens; it is perhaps futile even to try to evaluate its policies at a time when it is so necessary to act. The crucial questions today are: How can we save mankind and its cultural heritage? How can we guard Europe from further disaster?
>
> There can be no doubt that the present world crisis, and the suffering and privation which it has engendered, are in large measure responsible for the dangerous upheavals we witness today. In such times discontent breeds hatred, and hatred leads to acts of violence, revolution and even war. Thus, we see how distress and evil beget new distress and evil.
>
> Once again, as was the case twenty years ago, leading statesmen are faced with a tremendous responsibility. One can only hope that, before it is too late, they will devise for Europe the kind of international treaties and commitments whose meaning is so completely clear that all countries will come to view any attempt at warlike adventures as utterly futile. However, the work of statesmen can succeed only if they are backed by the sincere and determined will of the people . . .
>
> I should like to give expression to an idea which occurred to me recently. When I was living in solitude in the country, I noticed how the monotony of a quiet life stimulates the creative mind. There are certain occupations, even in modern society, which entail living in isolation and do not require great physical or intellectual effort. Such

> occupations as the service of lighthouses and lightships come to mind. Would it not be possible to place young people who wish to think about scientific problems, especially of a mathematical or philosophical nature, in such occupations? Very few young people with such ambitions have, even during the most productive period of their lives, the opportunity to devote themselves undisturbed for any length of time to problems of a scientific nature. Even if a young person is fortunate enough to obtain a scholarship for a limited period, he is pressured to arrive as soon as possible at definite conclusions. Such pressure can only be harmful to a student of pure science. In fact, the young scientist who enters a practical profession which earns him a livelihood is in a much better position, assuming, of course, that his profession affords him sufficient time and energy for his scientific work.

That address was Einstein's European valedictory.

Einstein's wife, his assistant, and Helen Dukas were already on board the *Westmoreland* when, on 7 October 1933, Albert joined them in Southampton.

Carrying visitors' visas, the four of them set out for a new life in the United States of America.

14

Arrival in the United States

1. Reception

On 17 October 1933, Einstein and his entourage arrived in New York, where they were met at quarantine by two trustees of the Institute, who handed him a letter from Flexner, the Institute's first director, which read in part:

> There is no doubt whatsoever that there are organized bands of irresponsible Nazis in this country. I have conferred with the local authorities . . . and the national government in Washington, and they have all given me the advice . . . that your safety in America depends upon silence and refraining from attendance at public functions . . . You and your wife will be thoroughly welcome at Princeton, but in the long run your safety will depend on your discretion.[238]

I must confess to some doubt as to whether the advice on discretion stemmed from Washington or from Mr Flexner, a Jew, himself. It is a sad chapter in Jewish history of the early parts of the twentieth century that Jewish migrations from East toward West, resulting from persecution, caused anxiety among 'Western' Jews: would those coming from the East not disturb relations between Westerners and their host country?

Even before Einstein's definitive departure from Europe, a group of patriotic American women had protested to the State Department the granting of a visa to a person whom they described as a communist. Einstein's response was acerbic and witty:

> Never before have I been so brusquely rejected by the fair sex, at least never by so many of its members at once!
>
> How right they are, those vigilant, civic-minded ladies! Why open one's door to someone who devours hard-boiled capitalists with as much appetite and gusto as the Cretan Minotaur devoured luscious Greek maidens in days gone by; one who is wicked enough to reject every kind of war, except the inexorable war with one's own spouse!
>
> Give ye therefore heed to your prudent and loyal womenfolk and remember that the Capitol of mighty Rome was once saved by the cackling of her faithful geese.[239]

Attacks continued, as witness a letter Einstein sent on 7 July 1933 to both *The New York Times* and the *Times* of London:

> I have received a copy of a circular issued by the Better America Federation, containing alleged photographs of me purporting to show that I am connected with the Third [Communist] International.
>
> I have never had anything to do with the Third International and never have been in Russia.
>
> Furthermore, it is manifest that the pictures purporting to be my photographs do not resemble me. The pictures are probably an attempted forgery inspired by political motives.

Much more important than Einstein's reactions were those of United States authorities. I reproduce two documents. The first is a concurrent Resolution adopted by the legislature of the State of New Jersey on 30 January 1933, a few months after Einstein's appointment to the Institute had been approved, but long before his immigration:

> WHEREAS, Announcement has been made that the New Institute of Advanced Studies, situated in the State of New Jersey, has invited the eminent scientist, Professor Albert Einstein, to become a member of its faculty; and
>
> WHEREAS, It is further announced that the said Professor Albert Einstein has accepted the invitation; and
>
> WHEREAS, The outstanding scientific ability of Professor Einstein is universally acknowledged and commended; and
>
> WHEREAS, The addition of the said Professor Albert Einstein to the faculty of a New Jersey institution of learning brings with it honor to the State of New Jersey and to its people; and
>
> WHEREAS, We, the members of the New Jersey Legislature, are cognizant of the high position of eminence occupied in the scientific world by the said Professor Albert Einstein and of the honor that is conferred upon our state by his accepting membership in the faculty of one of our institutions within the state; now therefore

> *Be it resolved*, That the Legislature of the State of New Jersey do, and it does hereby welcome the said eminent Professor Albert Einstein to the State of New Jersey, to his new scene of scientific activities; and
>
> *Be it further resolved*, That an invitation be extended to Professor Albert Einstein to address the members of the legislature in the very near future; and
>
> *Be it further resolved*, That a copy of this resolution be forwarded to the said Professor Albert Einstein and to the New Institute of Advanced Studies, in New Jersey.

The second document, found in the Congressional Records, 73rd Congress, 2nd Session, 28 March 1934, reads:

> IN THE HOUSE OF REPRESENTATIVES
>
> March 28, 1934
>
> MR. KENNEY introduced the following joint resolution; which was referred to the Committee on Immigration and Naturalization and ordered to be printed
>
> ---
>
> JOINT RESOLUTION
>
> To admit Albert Einstein to citizenship.
>
> Whereas Professor Albert Einstein has been accepted by the scientific world as a savant and a genius; and
>
> Whereas his activities as a humanitarian have placed him high in the regard of countless of his fellowmen; and
>
> Whereas he has publicly declared on many occasions to be a lover of the United States and an admirer of its Constitution; and
>
> Whereas the United States is known in the world as a 'haven of liberty and true civilization':
>
> Therefore be it
>
> 1 *Resolved by the Senate and House of Representatives*
> 2 *of the United States of America in Congress assembled,*
> 3 That Albert Einstein is hereby unconditionally admitted to
> 4 the character and privileges of a citizen of the United States.

Two months before this Joint Resolution was adopted, Einstein had received a letter from President Franklin Delano Roosevelt inviting him and his wife to the White House. They spent a night in the Franklin Room.

A few little souls had found Einstein's presence in the United States objectionable – but official America had given him the red carpet treatment. The Einstein charisma had not waned.

2. Settling down in Princeton

I return to 17 October 1933, the day on which Einstein arrived in New York. After having passed quarantine, his party was taken by special tug to the Battery. From there, they were driven directly to Princeton, where rooms at the Peacock Inn were waiting for them. A few days later, the Einsteins and Helen Dukas moved to a rented house at 2 Library Place. There they stayed until 1935, when Einstein bought the house at 112 Mercer Street from Mary Marden, paying for it in cash. In the autumn of that year, they moved in.

How did Einstein acquire the funds to pay for his house? He was not penniless on arrival but, remember, his Berlin bank account had been blocked. Helen Dukas has told me in confidence how the money was acquired. Now that all participants are gone, I feel relieved of my promise of silence.

It went like this. The economist Otto Nathan, a friend of Einstein from the Berlin days and himself an emigré living in New York City, had suggested that Einstein get hold of the needed dollars by selling one of his handwritten manuscripts. Einstein had agreed. Officials from the Morgan Library in New York were contacted who were more than willing to act as buyers. That is why an important manuscript on relativity theory, dating from 1912, is now present in the holdings of that fine Library.

The house on Mercer Street was to be Einstein's last home. In 1939, Mussolini's racial laws forced Einstein's sister, Maja to leave the small estate outside Florence which Einstein had bought for her and her husband Paul Winteler. Maja came to live with her brother in Princeton; Paul (a non-Jew) moved to Geneva.

Death struck in the early years. In 1934 Ilse Einstein, Albert's older stepdaughter, died in Paris after a painful illness. Thereupon Margot, her younger sister, joined her family in Princeton.

In May 1935 Einstein, his wife, and Margot sailed for Bermuda, in order to obtain immigrant visas upon reentry. That was Einstein's last trip outside the United States.

In 1936 Elsa, Einstein's wife, became gravely ill, largely the result (Helen Dukas believed) of the shock following the death of her older daughter Ilse. She died of heart disease on 20 December of that year.

In 1937 Hans Albert, Albert's older son, arrived in New York on the Holland–America liner *Veendam*. Albert was at the pier to welcome him. His son, 'construction engineer in Zürich, said it was his first visit to the United States. His wife and two children remained in Europe, pending his decision to make America his permanent home.' Albert refused to be interviewed. 'After all, private life is private life.'[240]

In the year of his death, Hans Albert remarked about his father, in a press interview: 'Probably the only project he ever gave up on was me. He tried to give me advice, but he soon discovered that I was too stubborn and that he was just wasting his time.'[241]

Shortly after arriving in the United States, Einstein gave Queen Elizabeth of Belgium his early impressions of Princeton: 'A quaint ceremonious village of puny demigods on stilts.'[242] A year and a half later he wrote to her again: 'I have locked myself into quite hopeless scientific problems – the more so since, as an elderly man, I have remained estranged from the society here.'[243]

Princeton, small, genteel, was not like the Berlin of the Weimar days, large, vibrant, and perverse. Even a man with a strong inner life like Einstein had to adjust himself to a new environment. He did, and very well. The more peaceful new life began to grow on him. There was music in the home. He found old friends and made new ones. He could be seen on Carnegie Lake in the small sailing boat he had bought, which had been christened *Tinnef* by Helen Dukas (Yiddish for 'cheaply made'). The name stuck. He never owned a car nor did he ever learn to drive. There were occasional trips to New York and to other cities. There were vacations, on the shores of Long Island or in the Adirondacks.

In 1936 Einstein took out his citizen papers. On 22 June 1940 he, Margot, and Dukas went to the State House in Trenton to be sworn in as United States citizens by that wonderful judge Philip Forman. (I cherish his memory; he inducted me too.) After the conclusion of the ceremony, Einstein participated in a broadcast sponsored by the United States Immigration and Naturalization Service, one in a series: 'I am an American'. Part of what he said:

> I do feel that, in America, the development of the individual and his creative powers is possible, and that, to me, is the most valuable asset in life. In some countries men have neither political rights nor the opportunity for free intellectual development. But for most Americans such a situation would be intolerable. . . . I gather from what I have seen of Americans since I came here that they are not suited, either by temperament or tradition, to live under a totalitarian system. I believe that many of them would find life not worth living under such circumstances. Hence, it is all the more important for them to see to it that these liberties be preserved and protected. . . .
>
> Intellectuals often lack the gift of impressing their audiences. Among the outstanding American statesmen, Woodrow Wilson probably provides the clearest example of an intellectual. Yet not even Wilson seems to have mastered the art of dealing with men. At first glance, his greatest contribution, the League of Nations, appears to have failed. Still, despite the fact that the League was crippled by his contemporaries and rejected by his own country, I have no doubt that Wilson's work will one day emerge in more effective form. Only then will the stature of that great innovator be fully recognized. . . .
>
> I believe that America will prove that democracy is not merely a form of government based on a sound Constitution but is, in fact, a way of life tied to a great tradition, the tradition of moral strength.

> Today more than ever, the fate of the human race depends upon the moral strength of human beings.[244]

At the time of Einstein's arrival, the Institute did not yet have its own building. He and other faculty members were given space in Princeton University's 'old' Fine Hall (now the Gest Institute of Oriental Studies). After 1939 they moved to the Institute's newly built Fuld Hall. His only official duty was to attend faculty meetings. This he did until his retirement at the age of 65, in 1944, and continued to do until early 1950. During all those years he was readily accessible to anyone who wished to discuss science with him.

I conclude this section with an Einstein anecdote which once again I owe to the late Helen Dukas.

During a speech by a high official at a major reception for Einstein, the honored guest took out his pen and started scribbling equations on the back of his program, oblivious to everything. The speech ended with a great flourish. Everybody stood up, clapping hands and turning to Einstein. Helen whispered to him that he had to get up, which he did. Unaware of the

From right to left: Meyer Weisgal, celebrated fund-raiser for the Weizmann Institute in Israel, Helen Dukas, Einstein, Abraham Pais, and Mrs Weisgal, in the living room on Mercer Street, late 1940s. (Author's private collection.)

fact that the ovation was for him, he clapped his hands, too, until Helen hurriedly told him that he was the one for whom the audience was cheering.

3. Political opinions, 1933–1935: Pacifism reconsidered.

During the years 1933–45 Einstein spoke out less on political issues than he had done before or would do again after the war. The reasons for this relative quietude are obvious. In the early years he was not yet a US citizen. When the war came there was only one issue: to win it. From 1933 until after the war he desisted from advocating world disarmament and conscientious objection.

When, in about 1928, Einstein had begun advocating refusal of military service, he had remarked right away that one should beware that the patient would not die before the cure, as might well be the case for oppressed African tribes (p. 174). After Hitler had come to power he had taken the same position, at least for the time being, in regard to European nations. His change of opinion, no longer promoting war resistance, in fact advocating rearmament of Western nations, had caused consternation and sharp criticism in diehard pacifist circles.

Also after arriving in America, Einstein held to his guns. In 1935 he expressed his position firmly and clearly:[245]

> I favor any measure which I consider likely to bring mankind closer to the goal of a supranational organization. Up to a few years ago, the refusal to bear arms by courageous and self-sacrificing individuals *was* such a measure; however, it can no longer be recommended as a course of action, at least not to European countries. As long as democratic governments of a similar character existed in the larger nations, and as long as none of these countries based their plans for the future on a policy of military aggression, the refusal of military service on the part of a fairly large number of citizens might well have induced the governments of these nations to become more favorably disposed toward the concept of international arbitration of conflicts between nations. Moreover, refusal of military service was likely to educate public opinion toward genuine pacifism and make obvious the unethical, immoral aspects of compulsory military service. In such a context, refusal of military service constituted a constructive policy.
>
> Today, however, you must recognize that several powerful nations make it impossible for their citizens to adopt an independent political position. These nations have succeeded in misleading their citizenry through an all-pervasive military organization and through dissemination of false information by means of an enslaved press, a centralized

radio service and a system of education oriented toward an aggressive foreign policy. In those countries, refusal of military service means martyrdom and death for those courageous enough to adopt such an attitude. On the other hand, in those countries which still respect the political rights of their citizens, refusal of military service is likely to weaken the ability of the healthy sections of the civilized world to resist aggression. Therefore, today, no intelligent person should support the policy of refusing military service, at least not in Europe, which is particularly endangered. Under present circumstances, I do not believe that passive resistance, even if carried out in the most heroic manner, is a constructive policy. Other times require other means, although the final goal remains unchanged.

These are the reasons why, in the present political context, a convinced pacifist must seek to promote his beliefs in ways different from those of earlier, more peaceful times. He must work for closer cooperation among the peaceful nations in order to minimize, as much as possible, the chances of success on the part of those countries whose adventurous policies are based on violence and brigandage. In particular, I am thinking of well-considered and sustained co-operation between the United States and the British Empire, including possibly France and Russia.

15

1933–9: Einstein's first American years in the limelight

After having settled in Princeton, Einstein of course rapidly returned to his beloved physics research. It is not the purpose of this essay to report on what he did in that respect, yet a general comment on these endeavors is in order.

During the years to be discussed in this chapter, Einstein's research went into three directions. First, there were further elaborations of what may be called conventional general relativity,* the theory he had formulated in 1916, which had brought him world fame in 1919 (Chapter 3). That work was good and memorable, though not as basically novel as his main discoveries of earlier years (as I already noted on p. 173). Secondly, he continued to pursue his dream of a unified field theory. That work is

* For cognoscenti: He worked on gravitational lenses, and on the problem of the motion of a particle in a gravitational field.

forgettable. Finally, in that period the 'EPR paper' appeared, a collaboration with Nathan Rosen and Boris Podolsky which deals with the foundations of quantum mechanics. There are a number of physicists who consider this a fundamental contribution to that subject. I am not one of those.

I may also recall here that only once after 1933 did Einstein leave the United States. That was in 1935 when he and his family sailed for Bermuda in order to obtain immigrant visas upon reentry.

I now turn to my main topic, to follow Einstein via the press from 1933 to the beginning of 1939, leaving for later, however, his comments on Jewish issues. We shall find that his activities became more socially oriented in those years, including not only interviews, but also after-dinner speeches, radio broadcasts, and commencement addresses.

1933. On 18 December Einstein spoke at a dinner in New York commemorating the hundredth anniversary of the birth of Alfred Nobel:

> We are gathered here today to express our gratitude to the magnanimous creator of the Nobel Foundation. We can perhaps accomplish this most effectively, I believe, by attempting to understand his [Nobel's] motivations. What may have happened inside this man to make him decide on so unique a last will?
>
> I believe the answer to this question may be found in the fact that the attainment of economic power is rarely based on productive ability or creative achievement. Inventiveness and organizational ability derive from altogether different talents and are seldom combined in one person. Schopenhauer was not wrong in considering the will and the intellect as mutually antagonistic.
>
> Probably, Nobel was primarily a creative mind. While his talent for organization predestined him to the attainment of dominating power, what really concerned him was the development of the creative spirit and the cultivation of a rich personality. Love of personal and intellectual freedom necessarily leads to a passionate preoccupation with the problem of peace since there exists no greater menace to personal freedom than war and militarism.
>
> That Nobel's chief creative achievements benefited the very powers which he considered most evil and destructive may well have caused him much distress. Thus we should view his testament as a heroic effort on his part to ensure that the fruits of his lifework would solve the painful contradiction in his personality. Thus the testament amounts to an act of the noblest self-emancipation.
>
> It is men like Nobel who will help us to find solutions to the burning social and economic problems that beset us today – men to whom economic achievements are but tools to serve the development of *human* values.[246]

1934. March. 'The cultural ideal of present-day America is not so much knowledge as it is the desire and the whole ability for accomplishment.' Einstein thinks that on the whole that is good – one carries not too much ballast that way.[247]

April. On getting acquainted with mathematics for the first time: 'It seemed to me a revelation of the Highest Author, and I will never forget it.'[248]

On religion: 'Organized religion may relieve some of the respect it lost in the last war if it dedicates itself to mobilizing the good will and energy of its followers against the rising tide of illiberalism.'[249]

In December Einstein spoke in Pittsburgh at the winter session of the American Association for the Advancement of Science. From a newspaper report: 'Atom energy hope is spiked by Einstein. Efforts at loosing vast force is called fruitless . . . It is something like shooting birds in the dark in a country where there are only few birds.'[250]

1935. On the occasion of anti-war demonstrations on several American campuses in the spring of 1935, Einstein made the following statement:

> It is a promising sign that America's academic youth addresses itself so zealously to the most important problem of the day, the preservation of peace. The creation of a serious attitude of good will is the first indispensable step toward this goal. The second step, equally indispensable, is a clear recognition of the means by which this goal may be attained . . .
>
> Do not be satisfied with merely expressing fine sentiments and pretty words about peace. You should approach the problem as a practical task of great magnitude and should lend your support to whatever organization seems to understand the problem most clearly and knows what methods should be employed toward its solution.[251]

In that year Einstein gave a number of after-dinner speeches in New York.

He spoke at the commemoration of the 800th anniversary of the birth of Maimonides.[252] From his script:[253]

> There is something sublime in the spectacle of man joining together in a spirit of harmony to honor the memory of a man whose life and work lie seven centuries in the past. This feeling is accentuated all the more sharply at a time in which passion and strife tend more than usually to obscure the influence of reasoned thought and balanced justice. In the bustle of everyday life our view grows clouded with desire and passion, and the voice of reason and justice is almost inaudible in the hubbub of the struggle of all against all. But the ferment of those times long past has long since been stilled, and scarcely more is left of it than the memory of those few who exerted a crucial and fruitful influence on their contemporaries and thus on later generations as well. Such a man was Maimonides.

Once the Teutonic barbarians had destroyed Europe's ancient culture, a new and finer cultural life slowly began to flow from two sources that had somehow escaped being altogether buried in the general havoc – the Jewish Bible and Greek philosophy and art. The union of these two sources, so different one from the other, marks the beginning of our present cultural epoch, and from that union, directly or indirectly, has sprung all that makes up the true values of our present-day life.

Maimonides was one of those strong personalities who by their writings and their human endeavors helped to bring about that synthesis, thus paving the way for later developments.

May this hour of grateful remembrance serve to strengthen within us the love and esteem in which we hold the treasures of our culture, gained in such bitter struggle. Our fight to preserve those treasures against the present powers of darkness and barbarism cannot then but carry the day.

Einstein also spoke (in German*) at a dinner given by the United Jewish Appeal,[254] and at a dinner honoring the theater director Max Reinhardt.[255] A further speech, this one given at a dinner organized by the American Christian Committee for German refugees, a few weeks after Italy's invasion of Ethiopia, was broadcast over a nation-wide radio network.[256] From Einstein's script:[257]

The process of cultural disintegration, which has assumed such dangerous proportions in Central Europe during the past few years, must alarm anyone who is sincerely interested in the welfare of humanity. Because neither the establishment of an international organization nor the sense of responsibility among nations has progressed sufficiently to allow joint action against the disease, efforts in two different directions are being made to protect our cultural values from the danger that threatens them.

The first and most important of these efforts must be an attempt to bring about a consolidation, within the framework of the League of Nations, of those nations which have not been directly affected by recent developments in Europe. Such a consolidation should have as its aim the common defense of peace and the establishment of military security. The second effort is to help those individuals who have been compelled to emigrate from Germany, either because their lives were endangered or their livelihood was taken away from them. The situation of these people is particularly precarious because of the economic crisis throughout the world and the high level of unemployment in almost all countries, which has frequently led to regulations prohibiting the employment of aliens.

It is well known that German fascism has been particularly violent in its attack upon my Jewish brothers. We have here the spectacle of the persecution of a group which constitutes a religious community. The

* I do not know what language Einstein used at other occasions in that period.

alleged reason for this persecution is the desire to purify the 'Aryan' race in Germany. As a matter of fact, no such 'Aryan' race exists; this fiction has been invented solely to justify the persecution and expropriation of the Jews.

The Jews of all countries have come to the assistance of their impoverished brothers as best they could and have also helped the non-Jewish victims of fascism. But the combined forces of the Jewish community have not nearly sufficed to help all these victims of Nazi terror. Hence, the emergency among the non-Jewish emigrants – that is to say, people of partly Jewish origin, liberals, socialists and pacifists, who are endangered because of their previous political activities or their refusal to comply with Nazi rules – is often even more serious than that of the Jewish refugees. . . .

To help these victims of Fascism constitutes an act of humanity, an attempt to save important cultural values and, not the least, a gesture of considerable political significance . . . To allow the condition of these victims to deteriorate further would not only be a heavy blow to all who believe in human solidarity but would encourage those who believe in force and oppression.

Einstein's most dramatic involvement with the press in 1935 concerned the case of Carl von Ossietzky.[258]

Ossietzky was chief editor of *Die Weltbühne*, a pacifist political weekly in Berlin, when on 12 March 1929, an article appeared in its columns in which it was revealed that much of the research and development for German civil aviation was secretly directed toward military purposes. Both the author of the article and Ossietzky were accused of treason and sentenced to eighteen months in jail. He received amnesty in December 1932. In February 1933, very soon after the Nazis had come to power, he was sent to a concentration camp.

Einstein then suggested to Jane Addams, an American winner of the Nobel Prize for peace, that she propose Ossietzky for that prize. That came to pass; in this effort she was joined by many noted European intellectuals. Despite the great caution of all involved, word leaked into the press. In May and June several New York dailies carried the story. In all these accounts Einstein was named.[259]

On 27 October 1935, Einstein himself wrote directly to the Nobel Committee in Oslo to propose Ossietzky, 'a man who, by his actions and his suffering, is more deserving of [the prize] than any other living person'. Such an award, he continued, would be 'a historic act that would suit to a high degree the solution of the peace problem'.

This is how this truly tragic story continued and ended. In January 1936, more than 500 members of the parliaments of Czechoslovakia, England, France, Holland, Norway, Sweden, and Switzerland signed petitions nominating Ossietzky for the peace prize. He stayed in the concentration camp until May 1936, when he was moved to a prison hospital with severe tuberculosis. In the autumn of 1936, Goering offered him freedom in exchange for a declaration that he would refuse the peace prize if it were awarded to him. Von Ossietzky refused. In November 1936 he was awarded the peace prize for 1935. On 30 January 1937, Hitler decreed that no German was henceforth permitted to receive Nobel prizes of any kind. The Nobel committee nevertheless awarded to Germans the chemistry prize in 1938 and the medicine prize in 1939. Both awards were declined. Von Ossietzky stayed in a prison hospital until, in May 1938, he died of tuberculosis. After his death Einstein wrote an eloquent and deeply felt short eulogy.[260]

1936. February. In London, Sir William Bragg, a noted British physicist, sitting in the chair of Michael Faraday, Britain's greatest nineteenth-century experimental physicist in the domain of electromagnetic phenomena, lit a candle which created a photoelectric impulse that was carried over the ocean and was used for dedicating the New York museum for science and industry in its new home at Rockefeller Center.

Einstein spoke at the opening in New York. 'Most [men] live like strangers in the world entrusted to them. . . . There is one advantage which primitive man has over civilized man – he is acquainted with the primitive tools he is using; he can make his own bow and arrow, in fact even his own canoe. How many civilized people have to any extent a clear idea about the nature and origin of the things they use for consumption, of which they take so much as a matter of course?'[261]

March. On playing games. 'I do not play games . . . There is no time for it. When I get through work I don't want anything that requires the working of the mind.' (I was pleased to read these comments, which so accurately describe my own sentiments, as they must for many scientists.) He had played chess once or twice as a boy but never indulged in bridge for relaxation. He had never heard of Monopoly. When the game was explained to him, he chuckled and remarked: 'A very American game'.[262]

Also in March. At a dinner in New York, Einstein handed out medals bestowed by the Einstein Award Foundation for 'service in fields of peace, literature, and humanity'. This press report includes a rare picture of Einstein wearing a tuxedo![263]

October. After having received an honorary degree from the State University of New York in Albany on the occasion of the celebration of the tercentenary of higher education in America, Einstein gave an address 'On Education',[264] part of which follows:[265]

> [I shall speak] about such questions as, independently of space and time, always have been and will be connected with educational matters.

In this attempt I cannot lay any claim to being an authority, especially as intelligent and well-meaning men of all times have dealt with educational problems and have certainly repeatedly expressed their views clearly about these matters. From what source shall I, as a partial layman in the realm of pedagogy, derive courage to expound opinions with no foundations except personal experience and personal conviction? If it were really a scientific matter, one would probably be tempted to silence by such considerations.

However, with the affairs of active human beings it is different. Here knowledge of truth alone does not suffice; on the contrary this knowledge must continually be renewed by ceaseless effort, if it is not to be lost. It resembles a statue of marble which stands in the desert and is continuously threatened with burial by the shifting sand. The hands of service must ever be at work, in order that the marble continue lastingly to shine in the sun. To these serving hands mine also shall belong.

The school has always been the most important means of transferring the wealth of tradition from one generation to the next. This applies today in an even higher degree than in former times, for through modern development of the economic life, the family as bearer of tradition and education has been weakened. The continuance and health of human society is therefore in a still higher degree dependent on the school than formerly.

The educational influence which is exercised upon the pupil by the accomplishment of one and the same work may be widely different, depending upon whether fear of hurt, egoistic passion, or desire for pleasure and satisfaction is at the bottom of this work. And nobody will maintain that the administration of the school and the attitude of the teachers do not have an influence upon the molding of the psychological foundation for pupils.

To me the worst thing seems to be for a school principally to work with methods of fear, force, and artificial authority. Such treatment destroys the sound sentiments, the sincerity, and the self-confidence of the pupil. It produces the submissive subject.

The second-named motive, ambition or, in milder terms, the aiming at recognition and consideration, lies firmly fixed in human nature. With absence of mental stimulus of this kind, human cooperation would be entirely impossible; the desire for the approval of one's fellow man certainly is one of the most important binding powers of society. In this complex of feelings, constructive and destructive forces lie closely together. Desire for approval and recognition is a healthy motive; but the desire to be acknowledged as better, stronger, or more intelligent than a fellow being or fellow scholar easily leads to an excessively egoistic psychological adjustment, which may become injurious for the individual and for the community. Therefore the school and the teacher must guard against employing the easy method of creating individual ambition, in order to induce the pupils to diligent work. One should guard against preaching to the young man success in the customary form as the main aim in life. The most important motive for work in the school and in life is the pleasure in work, pleasure in its result, and the knowledge of the value of the result to the community.

The point is to develop the childlike inclination for play and the childlike desire for recognition and to guide the child over to important fields for society. Such a school demands from the teacher that he be a kind of artist in his province.

I have said nothing yet about the choice of subjects for instruction, nor about the method of teaching. Should language predominate or technical education in science?

To this I answer: in my opinion all this is of secondary importance. If a young man has trained his muscles and physical endurance by gymnastics and walking, he will later be fitted for every physical work. This is also analogous to the training of the mind and the exercising of the mental and manual skill. Thus the wit was not wrong who defined education in this way: 'Education is that which remains, if one has forgotten everything he learned in school!'

Finally, in about 1936 Einstein wrote a brief but revealing self-portrait (this item is not from a press report):

Of what is significant in one's own existence one is hardly aware, and it certainly should not bother the other fellow. What does a fish know about the water in which he swims all his life?

The bitter and the sweet come from the outside, the hard from within, from one's own efforts. For the most part I do the thing which my own nature drives me to do. It is embarrassing to earn so much respect and love for it. Arrows of hate have been shot at me too; but they never hit me, because somehow they belonged to another world, with which I have no connection whatsoever.

I live in that solitude which is painful in youth, but delicious in the years of maturity.[266]

1937. On 18 July 1936 the Spanish Civil War had broken out, which was to end in April 1939 with a victory for Francisco Franco and his Fascist rebel forces. On 18 April 1937 a mass meeting was held in New York in support of the embattled Spanish Republic. Explaining that ill health kept him from attending in person, Einstein sent this message:

Let me emphasize above all that I view vigorous action to save freedom in Spain as the inescapable duty of all true democrats. Such a duty would also exist even if the Spanish Government and the Spanish people had not given such admirable proof of their courage and heroism. The loss of political freedom in Spain would seriously endanger political freedom in France, the birthplace of human rights. May you succeed in rousing the public to giving active support. . . . The success of your just and significant cause is very close to my heart.[267]

1938. Together with a group of Princeton University professors, Einstein signed an appeal to lift the US weapons embargo to the Spanish (i.e., anti-Franco) government. This news was widely reported in the daily papers. Among the press reactions, the following one, from a Brooklyn paper,[268] is strikingly anti-semitic:

SEND EINSTEIN BACK

A group of professors at Princeton University have appealed to the United States Government to drop the embargo on arms to Spain. The first signer was Professor Einstein, who recently fled here from Germany where the Jews are being persecuted.

Professor Einstein at a time of personal peril was given sanctuary in this land. Now he is engaged in telling our government how to run its business. This is sufficiently impertinent and arrogant but what is worse is to have him indorsing a move to shoot down and continue the persecution of Christians in Spain.

Einstein is a type. He seems to think that to his kind belong the world and the rest are put on it merely to be trampled over. In reference to persecution the only ones worthy of assistance are apparently the Jews, in fact the only ones capable of being persecuted are the Jews. If Christians are persecuted, the same is not to be recognized; the attitude seems to be 'what right have they to exist?'

Someone might say: 'Wouldn't you think as long as this country took Einstein in out of the storm he would at least wait a few years before dictating to the government?' That is incorrect. The Einsteins think every government, every country, is particularly theirs, that it is to be run for their benefit and the devil with the rest.

We believe the National Conference of Jews and Christians, the various interfaith and brotherhood movements, will be laughed out of favor unless they take up effectively the Gersons, Isaacs, Rabbi Stephen Wises, Einsteins, etc. etc. whose activities are causing more widespread anti-Semitism here than all Hitler's hirelings together. As suggestion number one toward good will, we suggest these good will societies recommend Einstein be sent back to Germany, where he may fully realize how to mind his own business and where persecution again might impress him with its heinousness in such a way that he will really hate it and hate it for Christians as well as Jews.

April. On April 1938 Einstein appeared for the first time on the cover of *Time* magazine. From the article in that issue:

On the bulletin board in Fine Hall [in Princeton] this elderly man was listed as 'A. Einstein', occupant of Room No. 215. . . . At present Einstein does not know whether the universe is finite or not. . . . The man who is regarded as the world's greatest living scientist lives placidly in a white frame house on Princeton's Mercer Street. He chose it for two dimensions, the height of its ceilings and the length of its flower garden in the back. He lives there with Margot, his late wife's daughter by a previous marriage, and his secretary, Fräulein Helen Dukas, who since Frau Einstein's death last year has looked after his bank account, his clothes and other things which to him are equally trivial. In the morning he works at home with his assistant, Dr. Peter G. Bergmann, a member of the Institute for Advanced Study. In the afternoons he goes to his office in Fine Hall. In the evenings he goes to concerts whenever possible, once in a long while to the cinema.

The Albert Einstein of today is no longer the timid bewildered man who visited the U.S. in 1930. He has acquired considerable poise in public, is not so afraid of the world as he used to be, entertains frequently. He has learned that it is not necessary to associate with anyone whom he does not like and trust. His telephone number is not listed and the telephone company will not furnish it. He leads the kind of life he likes and the U.S. suits him very well.

An accounting of Frau Elsa Einstein's estate filed last week in Trenton revealed that she left $52,689. Since she died intestate, only one-third will go to Widower Einstein, the rest to her daughter. This means nothing to Einstein. He has enough money for living expenses and wants no more.

June. On 6 June, Einstein delivered the commencement address, on 'Morals and Emotions', at Swarthmore College.[269] From his text:

Long before men were ripe to be faced with a universal moral attitude, fear of the dangers of life had led them to attribute to various imaginary personal beings, not physically tangible, power to release those natural forces which men feared or perhaps welcomed. And they believed that those beings, which everywhere dominated their imagination, were psychically made in their own image, but were endowed with superhuman powers. These were the primitive precursors of the idea of God. Sprung in the first place from the fears which filled man's daily life, the belief in the existence of such beings, and in their extraordinary powers, has had so strong an influence on men and their conduct, that it is difficult for us to imagine. Hence it is not surprising that those who set out to establish the moral idea, as embracing all men equally, did so by linking it closely with religion. And the fact that those moral claims were the same for all men, may have had much to do with the development of mankind's religious culture from polytheism to monotheism.

The universal moral idea thus owed its original psychological potency to that link with religion. Yet in another sense that close association was fatal for the moral idea. Monotheistic religion acquired different forms with various peoples and groups. Although those differences were by no means fundamental, yet they soon were felt more strongly than the essentials that were common. And in that way religion often caused enmity and conflict, instead of binding mankind together with the universal moral idea.

Then came the growth of the natural sciences, with their great influence on thought and practical life, weakening still more in modern times the religious sentiment of the peoples. The causal and objective mode of thinking – though not necessarily in contradiction with the religious sphere – leaves in most people little room for a deepening religious sense. And because of the traditional close link between religion and morals, that has brought with it, in the last hundred years or so, a serious weakening of moral thought and sentiment. That, to my mind, is a main cause for the barbarization of political ways in our time. Taken together with the terrifying efficiency of the new technical means, the barbarization already forms a fearful threat for the civilized world.

> Needless to say, one is glad that religion strives to work for the realization of the moral principle. Yet the moral imperative is not a matter for church and religion alone, but the most precious traditional possession of all mankind. Consider from this standpoint the position of the Press, or of the schools with their competitive method! Everything is dominated by the cult of efficiency and of success and not by the value of things and men in relation to the moral ends of human society. To that must be added the moral deterioration resulting from a ruthless economic struggle. The deliberate nurturing of the moral sense also outside the religious sphere, however, should help also in this, to lead men to look upon social problems as so many opportunities for joyous service towards a better life. For looked at from a simple human point of view, moral conduct does not mean merely a stern demand to renounce some of the desired joys of life, but rather a sociable interest in a happier lot for all men . . .
>
> Morality in the sense here briefly indicated is not a fixed and stark system. It is rather a standpoint from which all questions which arise in life could and should be judged. It is a task never finished, something always present to guide our judgment and inspire our conduct.

September. It was announced in the *New York Times*[270] that, at the initiative of the Westinghouse Company, a capsule would be deposited at the site of the coming New York World Fair. It was a short metal cocoon holding contemporary memorabilia, and is to be opened 5000 years later, on 23 September 6939. Included were messages by Einstein, Thomas Mann, and Millikan. Einstein's contribution:

> Our time is rich, in inventive minds, the inventions of which could facilitate our lives considerably. We are crossing the seas by power and utilize power also in order to relieve humanity from all tiring muscular work. We have learned to fly and we are able to send messages and news without any difficulty over the entire world through electric waves.
>
> However, the production and distribution of commodities is entirely unorganized so that everybody must live in fear of being eliminated from the economic cycle, in this way suffering for the want of everything. Furthermore, people living in different countries kill each other at irregular time intervals, so that also for this reason anyone who thinks about the future must live in fear and terror. This is due to the fact that the intelligence and character of the masses are incomparably lower than the intelligence and character of the few who produce something valuable for the community.
>
> I trust that posterity will read these statements with a feeling of proud and justified superiority.[271]

October. Einstein was one of the principal speakers at the dedication of the new refugee shelter of the American Jewish Congress at 48 West 68th Street in New York.[272]

1939. In January, Thomas Mann received the Einstein prize awarded by the Jewish Forum. The presentation was made by Einstein himself, who said on that occasion:

> The standard bearers have grown weak in the defense of their priceless heritage, and the powers of darkness have been strengthened thereby. Weakness of attitude becomes weakness of character; it becomes lack of power to act with courage proportionate to danger. All this must lead to the destruction of our intellectual life unless the danger summons up strong personalities able to fill the lukewarm and discouraged with new strength and resolution.[273]

In that month of January a scientific discovery was made that would turn out to be earth-shaking, literally.

16

About nuclear fission and atomic weapons; and, of course, more varia

1. The prehistory of the atomic bomb

In the autumn of 1938, Otto Hahn, and Fritz Strassman, two scientists from Berlin, began a careful radiochemical analysis of the elements produced in collisions of neutrons with uranium nuclei. Among the resulting products they identified barium, an element whose atoms weigh about half as much as a uranium atom.

It was a discovery that can be called staggering; nothing like it had ever been seen or dreamed of before. Until that time nuclear reactions had never produced elements that differed so much in weight from that of the bombarded nuclei. Those results appeared in print on 6 January 1939.[274]

The theoretical analysis of this phenomenon was given by two German-Jewish emigrés, Lise Meitner and Otto Frisch. In their resulting paper[275] they introduced a name for this new process. Frisch had asked a biologist what term he used for biological cell division. 'Fission', was the answer. Fission became the name also for the novel type of nuclear reaction.

Fission is a violent nuclear fragmentation in which a lot of energy is released and additional neutrons are produced. These neutrons can attack neighboring uranium nuclei, leading to further fission, and the release of further energy and neutrons. And on it goes. It did not take great wisdom to realize that this chain reaction might, perhaps, be a mechanism for

SECOND NEWS SECTION

For News, Subscriptions or Advertising Call ATlantic 6100.

SATURDAY MORNING,

Pittsburgh Post-Gazette

DECEMBER 29, 1934.

SPORTS, FINANCIAL, CLASSIFIED SECTION

This WORLD of OURS

by Clara Bride Brokaw

FOR MEN ONLY

Constantly Expressive—Einstein Warms Up to His Subject

ATOM ENERGY HOPE IS SPIKED BY EINSTEIN

Efforts at Loosing Vast Force Is Called Fruitless.

SAVANT TALKS HERE

Now Indicates Doubt Of Relativity Theory He Made Famous.

Energy of Atom.

Einstein talking to reporters about atomic energy, December 29, 1934. (Pittsburgh Post-Gazette. Courtesy AIP Emilio Segrè Visual Archives.)

producing extremely violent explosives. That possibility had already been publicly discussed in 1939 by physicists – and also in the press. Just one sample, a headline in the *Washington Post* of 29 April 1939: 'Physicists debate whether experiments will blow up 2 miles of the landscape.' As everybody now knows, the answer is 'yes'. So much for the prehistory of the atomic bomb.

As late as March 1939, however, Einstein had his doubts about the possibility of constructing such weapons. In an interview[276] on the occasion of his 60th birthday, he stated:

> The results gained thus far concerning the splitting of the atom do not justify the assumption that the atomic energy released in the process could be economically utilized. Yet, there can hardly be a physicist with so little intellectual curiosity that his interest in this important subject could become impaired because of the unfavorable conclusion to be drawn from past experimentation.

Within months he changed his mind, after having realized that he had overlooked a crucial aspect of the fission process. More about that presently, after I have first mentioned his public utterances during 1939 on other issues.

2. 1939, varia

May. The first illumination of the New York World Fair is initiated by cosmic rays. Einstein gives a five-minute address on the discovery of cosmic rays to an audience of 200,000 visitors.[277]

Also in May. 50,000 people attend the opening of the Jewish Palestine pavilion at the world's fair. Einstein, in his words, was 'entrusted with the high privilege of officially dedicating the building which my Jewish brethren have erected as their contribution to the world fair'.[278]

July. The film *World leaders on peace and democracy* has its première at the world fair. The leaders included Edvard Beneš, the exiled former president of Czechoslovakia, Arthur Compton, Einstein, Thomas Mann, and Urey. Einstein suggests that the real solution of the problem of peace will come only if nations accept 'a binding obligation to submit all cases in dispute to a super-court for decision, to accept these decisions under any circumstance, and to cooperate in carrying them out.'[279]

August. Einstein sends greetings to the World Student Organization meeting in Paris, attended by 200 delegates from 35 countries. The meeting was designed to bring anti-Fascist students closer together.[280]

November. Einstein formulates four basic axioms for education: 'Firmly establish certain moral and social principles and standards, and conduct the character education of youth along these lines; develop important intellectual capacities, such as logical thinking, judgment, memory, art appreciation, creative ability, as well as physical fitness; transmit general knowledge and information, as well as skill in routine functions such as reading, writing, arithmetic, and languages; impart special knowledge and skill in preparation for a profession.'[281]

3. Einstein and FDR

As noted above, in March 1939 Einstein did not believe in practical applications of atomic energy. By August he had changed his mind, as the result of discussions with Leo Szilard and Wigner, who at some time in July had visited him in his holiday retreat in Peconic on Long Island.

These two physicists had become concerned about the possibility that the Germans might produce atomic bombs and therefore felt that an American effort was urgently necessary. Szilard has recalled that when they explained to Einstein the possibility of a chain reaction, he exclaimed: '*Daran habe ich gar nicht gedacht*!' ('I never thought of that') '. . . [He was] very quick to see the implications and perfectly willing to do anything that needed to be done. He was quite willing to assume responsibility for sounding the alarm even though it was quite possible that the alarm might prove to be a false alarm. The one thing most scientists are really afraid of is to make fools of themselves. Einstein was free from such a fear and this above all is what made his position unique on this occasion.'[282]

The three men concluded that President Roosevelt should be informed. Einstein dictated a letter in German to Wigner, who has kindly provided me with a photocopy of his handwritten draft. Thereupon Wigner produced a translation into English which Einstein signed on 2 August. This document was brought to Dr Alexander Sachs, an economist and unofficial adviser to the President, who handed the document to Roosevelt on 3 October 1939.

That letter obviously did not reach the press. Because of its importance, I nevertheless reproduce here some of its key phrases:

> In the course of the last four months it has been made probable . . . that it may become possible to set up nuclear chain reactions in a large mass of uranium, by which vast amounts of power and large quantities of new radium-like elements would be generated. Now it appears almost certain that this could be achieved in the immediate future.
>
> This new phenomenon would also lead to the construction of bombs . . .
>
> In view of this situation you may think it desirable to have some permanent contact maintained between the Administration and the group of physicists working on chain reactions in America. . . .

> I understand that Germany has actually stopped the sale of uranium from the Czechoslovakian mines which she has taken over. That she should have taken such early action might perhaps be understood on the ground that the son of the German Under Secretary of State . . . is attached to the Kaiser Wilhelm Institut in Berlin, where some of the American work on uranium is now being repeated.[283]

One month later the Second World War broke out . . .

On 19 October Roosevelt replied:

> My dear Professor,
>
> I want to thank you for your recent letter and the most interesting and important enclosure.
>
> I found this data of such import that I have convened a board consisting of the head of the Bureau of Standards and a chosen representative of the Army and Navy to thoroughly investigate the possibilities of your suggestion regarding the element of uranium. . . .
>
> Please accept my sincere thanks.[284]

Roosevelt's board consisted of respectable but not distinguished men. The government's first allocation of funds for the study of fission was a pittance: $6000 for a full year! No wonder that months went by with little activity. It was therefore suggested to Einstein that he write a follow-up letter to be seen by the President – which indeed happened. From that letter, dated 7 March 1940:

> Last year, when I realized that results of national importance might arise out of the research on uranium, I thought it my duty to inform the Administration of this possibility . . .
>
> Since the outbreak of the war, interest in uranium has intensified in Germany. I have now learned that research there is carried out in great secrecy and that it has been extended to another of the Kaiser Wilhelm Institutes, the Institute of Physics.[285]

Much has been written about this Einstein–Roosevelt correspondence. Allegations sometimes made that this was the initial stimulus for the Manhattan Project are without substance; Roosevelt's appointment of a board in October 1939 did not result in a real governmental plan of action. In fact it was only in October 1941 that he decided to go ahead with full-scale atomic weapons development.* It was not until then that Secretary of War Henry Stimson heard about the project for the first time![287]

The account given above represents the sum and substance of Einstein's wartime involvement with the atomic bomb. He was excluded from its wartime development.[282] Helen Dukas has told me of Einstein's reaction

* As I have expatiated elsewhere,[286] The man primarily responsible for getting the American project going was Vannevar Bush.

to the first newspaper report of the bombing of Hiroshima. All he said then was '*Oj weh*' ('Woe is me').

In his later years Einstein himself said more than once that he regretted having signed the letters to Roosevelt. 'Had I known that the Germans would not succeed in producing an atomic bomb, I would not have lifted a finger.'[288]

A final note about Einstein's wartime thinking about war. In December 1944, Niels Bohr, then in Washington, received a letter from Einstein.[289] The latter was alarmed by visits from the physicist Otto Stern, who had said to him that after the war a secret arms race would develop that necessarily would lead to worse wars than ever before. Einstein proposed that leading scientists should suggest to the political leaders that military forces should be internationalized, and asked Bohr to join him. Bohr hastened to Einstein (also in December) to make clear to him that such a *démarche* might have the most deplorable consequences if anyone with confidential information were to participate. Einstein replied that he understood, that he himself would refrain from action, and that he would urge other colleagues to do likewise.

4. Varia, 1940–5

1940. March. The appointment of Bertrand Russell to the City College of New York had caused considerable controversy. This led Einstein to declare: 'Great spirits have always encountered violent opposition from mediocre minds. The mediocre mind is incapable of understanding the man who refuses to bow blindly to conventional prejudices and chooses instead to express his opinions courageously and honestly.'[290]

May. In a telegram to Roosevelt, Einstein and a number of other Princeton academics expressed their opposition to the neutrality stand of the American Association of Scientific Workers:

> The undersigned men of science, residing in Princeton, express their emphatic disagreement with the petition prepared by the American Association of Scientific Workers. We believe that the interests of the United States as well as those of civilization everywhere are placed in imminent danger by totalitarian aggression and that our best national defense consists in assistance to those forces now opposing this aggression.[291]

1941. June. Einstein was present at the dedication ceremony in Hightstown, New Jersey, of a farm for *Hechalutz* (pioneers for Palestine). 'The professor spent more than 15 minutes inspecting the chicken houses.'[292]

December. On the 7th the Japanese attacked Pearl Harbor. On that day Einstein dictated over the phone a 'Message for Germany' to a White House correspondent:

> This war is a struggle between those who adhere to the principles of slavery and oppression and those who believe in the right of self-determination both for individuals and for nations. Man must ask himself: Am I no more than a tool of the state? Or is the state merely an institution which maintains law and order among human beings? I believe the answer is that, in the last analysis, the only justifiable purpose of political institutions is to assure the unhindered development of the individual and his capacities.
>
> This is why I consider myself particularly fortunate to be an American. America is today the hope of all honorable men who respect the rights of their fellow men and who believe in the principles of freedom and justice.[293]

Also in December. In response to a reporter's question: 'In the Twenties, when no dictatorship existed I advocated that refusing [to wage war] would mean making war improper. As soon as conditions came about that in certain States there was so much coercion that this method could not be employed, I felt that it would weaken the less aggressive nations in the face of the more aggressive . . . We must strike hard and leaving the breaking to the other sides.'[294]

1942. Einstein addressed – by phone from his home where he lay ill – a dinner in New York of the Jewish Council for Russian War Relief. He declared that after the war 'Russia will give powerful and loyal cooperation to a workable scheme of international security, if other powers show equal seriousness and good will.'[295]

1944. During the War Einstein acted as occasional consultant for the Naval Bureau of Ordnance. In March, the *New York Times* quoted from *Star Shell*, a publication of the Bureau: 'He is working on the theory of explosions, seeking to determine what laws govern the more obscure waves of detonation.' The account added that Einstein would not have to get a Navy haircut or wear a uniform. 'He thinks better in his old windbreaker with his trousers rolled up.'[296]

In the same article, Einstein was quoted as saying:

> 'Why is it that nobody understands me and everybody likes me?'

May. In a statement to the National Wartime Conference of scientific, political, and white-collar organizations, Einstein called for all 'to fight for the establishment of a supranational force as a protection against fresh wars of aggression. I see [this] as the most important service which an organization of intellectual workers can perform in this historic moment . . . An organization of intellectual workers can have the greatest significance for society as a whole by influencing public opinion through publicity and education.'[297]

Einstein with Oppenheimer, late 1940s. (International Communication Agency. Courtesy AIP Emilio Segrè Visual Archives.)

Albert Einstein (early 1950s). (National Archives, U.S.A. Courtesy AIP Emilio Segrè Visual Archives.)

August. Einstein repudiates a biography *Einstein – An Intimate Story of a Great Man*, by Dimitri Marianoff, as written by a man not competent and without the moral right to do so.[298] (In 1930 Margot, Einstein's younger stepdaughter, had married Marianoff. In 1937 she obtained a divorce on grounds of desertion.)

October. Einstein urges the public to support FDR for a fourth term, as being most qualified to deal with world security and peace.[299]

1945. April. After Roosevelt's death, on the 12th, Einstein prepared a message for a memorial meeting in New York which contains these words: 'No matter when this man might have left us, we would have felt that we had suffered an irreplaceable loss. It is tragic that he did not live to lend his unique abilities to the task of finding a solution to the problem of international security. . . . All people of good will feel that with the death of Roosevelt they lost an old and dear friend. May he have a lasting influence on the hearts and minds of men!'[300]

August. During the week following the bombing of Hiroshima, Einstein declared: 'I have done no work on [the atomic bomb], no work at all. I am interested in the bomb the same as any other person; perhaps a little more interested.'[301]

17

The final decade. Einstein and the Atomic Age

This chapter deals with Einstein's last few years, from the end of the Second World War until his death. I shall first relate some events in his personal life as noticed by the press, then, for the last time, present a menu of varia, and finally deal with his political statements in that period regarding atomic weapons.

1. Events of a personal nature: state of health; birthdays; unified field theory; presidency of Israel; honors and awards

In his last years, Einstein was not well for much of the time, suffering from attacks of pain in the upper abdomen. In the autumn of 1948 his surgeon diagnosed an abdominal growth the size of a grapefruit and suggested an exploratory laparotomy, to which Einstein consented. He entered the Jewish Hospital in Brooklyn where, on 31 December at 8 a.m., he was operated upon.[302] His surgeon discovered an aneurysm in the abdominal aorta, a growth that in those days could not yet be removed by an operation which by now is fairly routine. Einstein stayed in the hospital until the incision was sufficiently healed and returned home on 13 January 1949.[303] When leaving the hospital, he was set upon by press photographers whom he waved off: 'No! no! no! – I'll stay here all evening if necessary.'[304]

Monday 14 March 1949 was Einstein's seventieth birthday. His secretary commented to the press: 'Doctor Einstein is not celebrating anything today – he never does celebrate.'[305] (On his next birthday he said: 'Such celebrations are for children.'[306]) He did receive visitors, however – youngsters from the Reception Shelter of United Service for New Americans.[307] He also received an honorary degree from the Hebrew University,[308] 'a token of esteem that fills me with pure joy'.[309] On 19 March a symposium in his honor was held at Palmer Laboratory, Princeton University. There were scientific talks and also a recorded tribute by Niels Bohr was played.[310]

I was present at that symposium. Most of us were in our seats when Einstein arrived. My most vivid memory of that day is the brief hush before we all stood to greet him.

In 1953: 'After having refused to lend his name to any enterprise whatsoever, Dr. Einstein consented on his seventy-fourth birthday to have Yeshiva University's medical school [in New York] called the Albert Einstein College of Medicine.'[311] In honor of this event a luncheon was held at the Princeton Inn (which I attended). It was one of the rare occasions where Einstein appeared dressed up. 'His gray suit was well pressed and he wore a soft-collared tan shirt with a gray-and-black striped tie.'[312]

Einstein's last birthday to receive press coverage was his 75th, in 1954. There were reports of greetings from Jawaharlal Nehru, Prime Minister of India, Thomas Mann, and Bertrand Russell;[313] and from President Truman, the presidents of Italy and Israel, and Mrs Roosevelt.[314] In Jerusalem the President of the Hebrew University delivered an address in his honor; West German newspapers, writing with reverence, gave much space to the event.[313]

The Emergency Civil Liberties Committee, meeting in Princeton, wanted to bring flowers to his home, which he refused: 'You may bring flowers to my door when the last witch hunter is sentenced, but not before.'[314] From an article on Einstein in the magazine section of the *New York Times* of 14 March: The Institute bus drives him from his home to his office and back. 'I'd like to walk both ways but I can't do that now.' The driver asks if it is going to snow. '*He* hasn't told me yet.' His assistant tells: 'His patience is wonderful. He is always optimistic about the problem – even optimistic about failures. His attitude is "Well, we have learned something."'

During this final decade Einstein worked on his last version of a unified field theory.* The press took considerable notice. One day in December 1949 the *New York Times* came out with a front-page article on the theory under the heading: 'New Einstein theory gives a master key to the Universe.'[315] The article includes the reproduction of a page of Einstein's latest manuscript, which was about to be published by Princeton University Press as an appendix to a new edition of Einstein's wonderful little book *The Meaning of Relativity.* (I own the galley proofs of that article, which Einstein had given me for critical comment.) The Princeton Press announced their new publication at the December meeting of the American Association for the Advancement of Science, held in the Statler Hotel in New York. Einstein refused to be interviewed on his work and asked his secretary to relay this message to reporters: 'Come back and see me in twenty years.'[316]

A few days later, the pastor of the South Reformed Church, Fourth Avenue at 55th Street, New York, praised[317] the new work as a 'restatement of the ancient monotheism anticipated by Moses . . . The Einstein revelation at Christmas time is another manifestation of God's wonder working.' Oh, well! The rabbi of Congregation B'nai Jeshurun, 257 West 88th Street, New York, listed Einstein as the first among the ten greatest Jews of the past 50 years for 'revolutionizing the science of physics'.[318] The *New York Times* editorialized: 'It would be a bold man who would predict how long it will be before this newest Einsteinian theory is generally understood.'[319]

In February 1950, Einstein is reported to have said to a newspaper man that his new theory 'gave him satisfaction similar to the one he had when he first worked out the theory of relativity.'[320] In May he donated the original manuscript of his new theory to the Hebrew University on the occasion of its 25th anniversary. At that ceremony he said: 'The support for cultural life is of primary concern to the Jewish people. We would not be in existence today as a people without this continued activity in learning.'[321]

In March 1953 there was again a front-page article in the *New York Times* on the new theory. It is 'published today by the Princeton University Press as an appendix to the fourth edition of his famous book originally published in 1922.' Einstein declared that 'his concept of 1950 left one serious difficulty to be solved . . . This last step in the theory has been fully overcome in the last few months.'[322]

To my knowledge, that was the last time Einstein's theories made headlines. I find these final statements on Einstein's post-Second World

* For experts: it is a theory in which both the fundamental tensor and the connection are non-symmetrical.

War research very saddening, since this concluding chapter of his *oeuvre* is without merit. It is not a worthy epitaph to a scientific corpus that will be justly revered for all time.

On 19 November 1952 the front page of the *New York Times* showed this item:

> Professor Albert Einstein has informed Prime Minister David Ben-Gurion that he is unable to accept the Presidency of Israel . . . [Einstein] remained incommunicado throughout the evening [of November 18]. His secretary said that all questions on the subject would have to be made to the Israeli Embassy in Washington.[323]

What was that all about?

Chaim Weizmann, the first President of Israel, had died on 9 November 1952. Thereupon the Israeli government decided to offer the presidency to Einstein. On 16 November the Prime Minister sent the following urgent cable to the Ambassador in Washington:

> Please inquire immediately of Einstein whether he is prepared to become President of Israel if elected [by the Parliament]. Immediately after election he would have to come to Israel and become an Israeli citizen. He could continue with his scientific work without interference. Please cable his answer immediately. Ben Gurion.[324]

In fact, Einstein first heard this news one afternoon from the *New York Times*. What happened next has been described by a friend who was with Einstein that evening.

> About nine o'clock a telegram was delivered . . . from the Israeli ambassador in Washington, Mr. Abba Eban. The highly elaborate terms of the telegram . . . made it quite plain that the earlier report must be true, and the little quiet household was much ruffled. 'This is very awkward, very awkward,' the old gentleman was explaining while walking up and down in a state of agitation which was very unusual with him. He was not thinking of himself but of how to spare the Ambassador and the Israeli government embarrassment from his inevitable refusal. . . . He decided not to reply by telegram but to call Washington at once. [He got] through to the Ambassador, to whom he spoke briefly and almost humbly made plain his position.[325]

Honors and awards continued to pile up. I mention a few.

In 1946 Einstein received an honorary doctorate* from Lincoln University,

* He also received honorary degrees from Geneva, Zürich, Rostock, Madrid, Brussels, Buenos Aires, the Sorbonne, London, Oxford, Cambridge, Glasgow, Leeds, Manchester, Harvard, Princeton, New York State at Albany, and Yeshiva. This list is most probably incomplete. Einstein's favorite reaction to such honors was: '*Der Teufel scheisst auf den groszen Haufen*,' the devil craps on the big pile.

a school for blacks. Addressing the student body, Einstein expressed the belief that there was a 'great future' for the Negro and urged the assembled students to work long and hard and with lasting patience.[326]

In 1948 Einstein received the One World Award.[327] From his address given in Carnegie Hall, New York, on that occasion:

> In the course of my long life I have received from my fellow-men far more recognition than I deserve, and I confess that my sense of shame has always outweighed my pleasure therein. But never, on any previous occasion, has the pain so far outweighed the pleasure as now. For all of us who are concerned for peace and the triumph of reason and justice must today be keenly aware how small an influence reason and honest good-will exert upon events in the political field. But however that may be, and whatever fate may have in store for us, yet we may rest assured that without the tireless efforts of those who are concerned with the welfare of humanity as a whole, the lot of mankind would be still worse than in fact it even now is.[328]

In 1953, Einstein received the Lord and Taylor Award.[329] His tape-recorded message of thanks, broadcast by radio, included these words:

> It gives me great pleasure, indeed, to see the stubbornness of an incorrigible nonconformist warmly acclaimed. To be sure, we are concerned here with nonconformism in a remote field of endeavor, and no Senatorial committee has as yet felt impelled to tackle the important task of combating, also in this field, the dangers which threaten the inner security of the uncritical or else intimidated citizen.
>
> As for the words of warm praise addressed to me, I shall carefully refrain from disputing them. For who still believes that there is such a thing as genuine modesty? I should run the risk of being taken for just an old hypocrite. You will surely understand that I do not find the courage to brave this danger.
>
> Thus all that remains is to assure you of my gratitude.[330]

In a message to the Decalogue Society of Lawyers on the occasion of receiving, in 1954, their award of merit, Einstein said:

> The fear of communism has led to practices which have become incomprehensible to the rest of civilized mankind and expose our country to ridicule . . . The existence and validity of human rights are not written in the stars . . . A large part of history is . . . replete with the struggle for . . . human rights, an eternal struggle in which a final victory can never be won. But to tire in that struggle would mean the ruin of society.[331]

I finally note that the distinguished figures calling on Einstein at his home on Mercer Street included Nehru (in 1949[332]) and Ben Gurion (in 1951[333]).

2. Last varia

1946. 1 July. *Time* magazine has Einstein on the cover once again. The picture shows him against a mushroom cloud in the background. Inside the mushroom: $E=mc^2$. Caption: 'Cosmoclast Einstein. All matter is speed and flame.' From the text:

> Through the incomparable blast and flame that will follow, there will be dimly discernible, to those who are interested in cause and effect in history, the features of a shy, almost saintly, childlike little man with the soft brown eyes, the drooping facial lines of a world-weary hound, and hair like an aurora borealis . . . Einstein was the father of the bomb in two important ways: (1) it was his initiative which started U.S. bomb research; (2) it was his equation ($E=mc^2$) which made the atomic bomb theoretically possible.

As the reader of the preceding will know by now, both these reasons for Einstein's father-of-the-bomb image are fabricated.

September. Einstein writes to President Truman: 'Security against lynching is one of the most urgent tasks of our generation.'[334]

Two days later: Bias against the Negro is 'the worst disease from which the society of our nation suffers'.[335]

1947. October. 'The fate of the surviving victims of German persecution bears witness to the degree to which the moral conscience of mankind has been weakened.'[335a]

1948. Einstein includes Henry Wallace together with Roosevelt and Wendell Wilkie in the category of men 'who are above the petty bickering of the day and without any selfish interest.'[336]

1949. March. 'It is nothing short of a miracle that modern methods of instruction have not yet entirely strangled the holy curiosity of inquiry.'[337]

August. Einstein urges the establishment of an international center of sociological studies, 'where the ways and means would be studied to establish a better understanding among nations . . . without creating the obsession of the past, as is so often the case.'[338]

1950. Einstein accepts life membership in the Montreal Pipe Smokers Club. 'Pipe smoking contributes to a somewhat calm and objective judgment of human affairs.'[339]

1951. January. Einstein and Truman send messages to the New York Society for Ethical Culture on the occasion of its 75th anniversary.[340] Also that month, Einstein accepts the score of a symphony dedicated to him by an Israeli composer.[341]

February. Einstein, and also the mathematicians John von Neumann and Norbert Wiener, have their brain waves recorded 'while they were thinking and also while at rest. They generally showed differences from the average.'[342] The article includes a picture of Einstein lying down with electrodes attached to his head.

March. In Princeton, Einstein hands the first Einstein prize to the physicist Julian Schwinger and the logician Kurt Gödel. Einstein was also the chairman of the award committee for the new prize, established by Lewis Strauss. Einstein to Schwinger: 'You have deserved it'; to Gödel: 'You don't need it.'[343] (I attended the festivities.)

August. The Revd Michael Walsh, President of the Eastern States Division of the American Association of Jesuit Scientists, criticizes Einstein, 'who uses the prestige of his scientific positions to speak on non-scientific subjects.'[344]

1952. Einstein sponsors the Jewish Agency for Palestine's Israel Summer Institute, a vacation-time work-study tour of Israel designed for American students and educators.[345]

April. Einstein receives Viscount Herbert Samuel at his home, to discuss scientific differences. Lord Samuel continues to hold the impression, Einstein said, that 'contemporary physics are based on concepts somewhat analogous to the "smile of the absent cat"'.[346]

May. (Front page!) Johanna Mankiewicz, daughter of the Hollywood film writer, writes from Los Angeles for help in a geometry problem. Einstein's reply is shown.[347] Johanna's teacher rebuked her. 'Great scientists should be bothered only by great problems.'[348]

August. Graduates of Yeshiva University who have won National Science Foundation fellowships for postgraduate work in the sciences are received by Einstein.[349]

October. In an interview on education:

> It is not enough to teach a man a specialty. Through it he may become a kind of useful machine but not a harmoniously developed personality [The student] must acquire an understanding of and a lively feeling for values. He must acquire a vivid sense of the beautiful and of the morally good. Otherwise he more closely resembles a well-trained dog. These precious things are conveyed to the younger generation through personal contact with those who teach, not through textbooks. It is this that primarily constitutes and preserves culture . . . Overemphasis on the competitive system and premature specialization on the ground of immediate usefulness kills the spirit on which all cultural life depends, specialized knowledge included. . . . Independent critical thinking [should] be developed in the young human being, a development that is greatly jeopardized by overburdening with too many and too varied subjects. Overburdening necessarily leads to superficiality.[350]

Also in October. 'Dr. Albert Einstein declared that he would cast his vote in the Presidential election for Governor Adlai Stevenson.'[351]

November. Front page: 'Professor Einstein concedes that even he can err.' This during a two-hour testimony on a patent infringement suit brought by his friend the medical doctor Gustav Bucky from New York. Einstein said: 'I may have made a mistake in answering a pre-trial deposition question.' At issue were patent rights to a highly technical medical camera.[352]

1953. January. Yuri Zhandov, member of the Communist Party central committee and chief of its science section, had published an article in *Pravda* about a meeting in Moscow of the Society for the Dissemination of Political and Scientific Knowledge.

> Mr. Zhandov denounced as prime example of bourgeois reactionary tendencies in science the physicists Albert Einstein and Niels Bohr and the late British astronomer Sir Arthur Eddington. He said that Soviet scientists who had accepted their views and circulated them within Russia were guilty of allowing violent enemies of Marxism to circulate their opinions in Soviet society.[353]

That was written in the year Stalin died. One year later, '*Pravda* placed its support behind Einsteinian physics. The Communist party paper sharply rebuked Soviet scientists who sought to differentiate between the great physicist's scientific contributions and what is called his 'confused' philosophy. The article cited Lenin as having profoundly respected Doctor Einstein's scientific contributions.'[353a]

May. In a plea of support for the Hebrew University, Einstein said: 'For the young state to achieve real independence and conserve it, there must be a group of intellectuals and experts produced in the country itself.'[354]

October. In a sermon, Rabbi Israel Goldstein, President of the American Jewish Congress, suggests that a group of twelve Christians and Jewish philosophers meet in 'a small neutral country' to arrive at a solution of today's international problems. Einstein topped his list.[355]

December. Einstein is elected honorary president of the Hebrew University.[356]

3. 'The war is won, but the peace is not': Einstein on atomic weapons

I do not know who invented the myth that Einstein's equation $E = mc^2$ made the atomic bomb possible. It is true that this equation plays an important role in nuclear physics, but to say that this made possible the construction of weapons is like saying that the invention of the alphabet caused the Bible to be written.

However that may be, Einstein's views on the new weapons were eagerly sought after, and he was quite ready to state publicly his opinions on the drastic change, after the Second World War, which these gadgets had wrought in the balance of world power. Thus, in September 1945, he stated in a press interview[357] that only a world government could save the human race, that only such an institution could avoid new world wars from breaking out. In October he was a cosignatory of a letter to the *New York Times*[358] which opens eloquently:

> The first atomic bomb destroyed more than the city of Hiroshima. It also exploded our inherited, outdated political ideas.

Also in October, the first public post-war attack on Einstein was made, by an ultraconservative member of the U.S. House of Representatives: 'This foreign-born agitator would have us plunge into another European war in order to further the spread of Communism throughout the world, etc., etc.'[359]

In November, Einstein gave a lengthy interview to a prominent American monthly, from which I quote.[360]

> I do not believe that civilization will be wiped out in a war fought with the atomic bomb. Perhaps two-thirds of the people of the earth might be killed. But enough men capable of thinking, and enough books, would be left to start again, and civilization could be restored.
>
> I do not believe that the secret of the bomb should be given to the United Nations Organization. I do not believe it should be given to the Soviet Union. Either course would be like a man with capital, and wishing another man to work with him on some enterprise, starting out by simply giving that man half of his money. The other man might choose to start a rival enterprise, when what is wanted is his cooperation. The secret of the bomb should be committed to a world government, and the United States should immediately announce its readiness to give it to a world government. . . .
>
> Since the United States and Great Britain have the secret of the atomic bomb and the Soviet Union does not, they should invite the Soviet Union to prepare and present the first draft of a constitution of the proposed world government. That will help dispel the distrust of the Russians, which they already feel because the bomb is being kept a secret chiefly to prevent their having it. . . .
>
> 'Do I fear the tyranny of a world government? Of course I do. But I fear still more the coming of another war or wars. Any government is certain to be evil to some extent. But a world government is preferable to the far greater evil of wars, particularly with their intensified destructiveness.
>
> Now we have the atomic secret we must not lose it, and that is what we should risk doing if we give it to the United Nations Organization or to the Soviet Union. But we must make it clear as quickly as possible that we are not keeping the bomb a secret for the sake of our power,

> but in the hope of establishing peace through a world government, and we will do our utmost to bring this world government into being . . .
>
> We shall not have the secret very long. I know it is argued that no other country has money enough to spend on the development of the atomic bomb, which assures us the secret for a long time. It is a mistake often made in this country to measure things by the amount of money they cost. But other countries which have the materials and the men and care to apply them to the work of developing atomic power can do so, for men and materials and the decision to use them, and not money, are all that are needed.
>
> *I do not consider myself the father of the release of atomic energy* (my italics). My part in it was quite indirect. I did not, in fact foresee that it would be released in my time. I believed only that it was theoretically possible.

In December Einstein addressed the fifth Nobel anniversary dinner at the Hotel Astor in New York.[361] He said in part:

> The war is won, but the peace is not. The great powers, united in fighting, are now divided over the peace settlements. The world was promised freedom from fear, but in fact fear has increased tremendously since the termination of the war. The world was promised freedom from want, but large parts of the world are faced with starvation while others are living in abundance. The nations were promised liberation and justice. But we have witnessed, and are witnessing even now, the sad spectacle of 'liberating' armies firing into populations who want their independence and social equality, and supporting in those countries, by force of arms, such parties and personalities as appear to be most suited to serve vested interests. Territorial questions and arguments of power, obsolete though they are, still prevail over the essential demands of common welfare and justice. . . .
>
> The picture of our postwar world is not bright. So far as we, the physicists, are concerned, we are no politicians and it has never been our wish to meddle in politics. But we know a few things that the politicians do not know. And we feel the duty to speak up and to remind those responsible that there is no escape into easy comforts, there is no distance ahead for proceeding little by little and delaying the necessary changes into an indefinite future, there is no time left for petty bargaining. The situation calls for a courageous effort, for a radical change in our whole attitude, in the entire political concept.[362]

1946. February. 'I do not believe that civilization will be wiped out in a war fought with the atomic bomb. Perhaps two-thirds of the people of the earth might be killed but enough men, capable of thinking, and enough books would be left to start again, and civilization could be restored.'[363] (Here Einstein repeats himself.)

May. In a coast-to-coast broadcast over the ABC network, Einstein addressed the solution of the problem of international peace.[364]

> It is linked solely to an agreement on a grand scale between this country and Russia. . . . If such an agreement could be achieved, then these two powers alone would be able to cause the other nations to give up their sovereignty to the degree necessary for the establishment of military security for all.
>
> Now many will say that fundamental agreement with Russia is impossible under the present circumstances. Such a statement would be justified if the United States had made a serious attempt in this direction during the last year. I find, however, that the opposite has happened. . . . There was no need to manufacture new atomic bombs without letup and to appropriate 12 billion dollars for defense in a year in which no military threat was to be expected for the near future. It is senseless to recount here all the details, which all show that nothing has been done in order to alleviate Russia's distrust, a distrust which can very well be understood in the light of the events of the last decade and to whose origin we contributed not a little.
>
> . . . The only hope for protection lies in the securing of peace in a supranational way. . . . A world government must be created which is able to solve conflicts between nations by judicial decision. . . . This government must be based on a clear-cut constitution which is approved by the governments and the nations and which has the sole disposition of offensive weapons. . . .
>
> A person or a nation can be considered peace loving only if it is ready to cede its military force to the international authorities and to renounce every attempt or even means of achieving its interests abroad by the use of force. . . . The U.N. as it stands today has neither the military force nor the legal basis to bring about a state of international security. Nor does it take account of the actual distribution of power.

June. From a long interview published in the Sunday magazine section of the *New York Times*:[365]

> Many persons in other countries now regard America with great suspicion, not only because of the bomb but because they fear she will become imperialistic. I was sometimes not quite free from such fears myself. Although other countries might not fear Americans if they knew us as we know one another, as honest and sober neighbors, they also know that even a sober nation can become drunk with victory. If Germany had not been victorious in 1870, what tragedy for the human race might have been averted!
>
> We are still making bombs, and the bombs are making hate and suspicion. We are keeping secrets and secrets breed distrust. I do not say we should now turn loose the secret of the bomb; but are we ardently seeking a world in which there will be no need for bombs or secrets, a world in which science and men will be free?
>
> So long as we distrust Russia's secrecy and she distrusts ours, so long will we walk together to certain doom . . .
>
> Science has brought forth this danger, but the real problem is in the minds and hearts of men. We will not change the hearts of other men by mechanical devices; rather we must change *our own* hearts and

> speak bravely. We must be generous in giving the rest of the world the knowledge we have of the forces of nature, after establishing safeguards against possible abuse. We must not merely be willing, but must be actively eager to submit ourselves to the binding authority necessary for world security. We must realize we cannot simultaneously plan for war and for peace.
>
> When we are clear in heart and mind – only then shall we find courage to surmount the fear which haunts the world.

August. On the 2nd, the Emergency Committee of Atomic Scientists was formally incorporated in the State of New Jersey. Its purpose was to promote information to the general public about the politics of atomic weapons and about peaceful applications. Einstein was the chairman of this eight-man group. In a nationwide broadcast, he appealed for donations to help enlighten the populace. In his speech he said:[366]

> The intellectual workers cannot successfully intervene directly in the political struggle. They can achieve, however, the spreading of clear ideas about the situation and the possibility of successful action. They can contribute through enlightenment to prevent able statesmen from being hampered in their work by antiquated opinions and prejudices.

The Emergency Committee became inactive in January 1949.

Also in August, Einstein 'was sure that President Roosevelt would have forbidden the atomic bombing of Hiroshima had he been alive'.[367]

1947. On 1 August 1946 President Truman had signed a bill creating the US Atomic Energy Commission, to be supervised by a board of civilians. His nomination of David Lilienthal as chairman of the Board had met with considerable Senate opposition. In February, Einstein joined backers who endorsed Lilienthal.[367a]

July. On the eve of the second anniversary of the experimental explosion of the first atomic bomb, in New Mexico, Einstein pleaded for continued faith in the United Nations.[368]

Summer. Einstein lets the U.S. military have it:

> I must frankly confess that the foreign policy of the United States since the termination of hostilities has reminded me, sometimes irresistibly, of the attitude of Germany under Kaiser Wilhelm II, and I know that, independent of me, this analogy has most painfully occurred to others as well. It is characteristic of the military mentality that non-human factors (atomic bombs, strategic bases, weapons of all sorts, the possession of raw materials, etc.) are held essential, while the human being, his desires and thoughts – in short, the psychological factors – are considered as unimportant and secondary. Herein lies a certain resemblance to Marxism, at least insofar as its theoretical side alone is

kept in view. The individual is degraded to a mere instrument; he becomes 'human *matériel*'. The normal ends of human aspiration vanish with such a viewpoint. Instead, the military mentality raises 'naked power' as a goal in itself – one of the strangest illusions to which men can succumb.

In our time the military mentality is still more dangerous than formerly because the offensive weapons have become much more powerful than the defensive ones. Therefore it leads, by necessity, to preventive war. . . .

I see no other way out of prevailing conditions than a far-seeing, honest, and courageous policy with the aim of establishing security on supranational foundations.[369]

September. Einstein sends an Open Letter to the General Assembly of the United Nations.[370] From its text:

The delegates of fifty-five governments, meeting in the second General Assembly of the United Nations, undoubtedly will be aware of the fact that during the last two years – since the victory over the Axis powers – no appreciable progress has been made either toward the prevention of war or toward agreement in specific fields such as control of atomic energy and economic cooperation in the reconstruction of war-devastated areas.

The present impasse lies in the fact that there is no sufficient, reliable supranational authority. Thus the responsible leaders of all governments are obliged to act on the assumption of eventual war. Every step motivated by that assumption contributes to the general fear and distrust and hastens the final catastrophe. However strong national armaments may be, they do not create military security for any nation nor do they guarantee the maintenance of peace. . . .

The time has come for the U.N. to strengthen its moral authority by bold decisions. First, the authority of the General Assembly must be increased so that the Security Council as well as all other bodies of the U.N. will be subordinated to it. As long as there is a conflict of authority between the Assembly and the Security Council, the effectiveness of the whole institution will remain necessarily impaired.

We must assume that despite all efforts Russia and her allies may still find it advisable to stay out of such a world government. In that case – and only after all efforts have been made in utmost sincerity to obtain the cooperation of Russia and her allies – the other countries would have to proceed alone. . . .

Such a partial world government should make it clear from the beginning that its doors remain wide open to any non-member – particularly Russia – for participation on the basis of complete equality.[371]

November. Another long interview, published in a monthly magazine.[372]

Since the completion of the first atomic bomb nothing has been accomplished to make the world more safe from war, while much has been done to increase the destructiveness of war. I am not able to speak from any firsthand knowledge about the development of the atomic bomb, since I do not work in this field. But enough has been said by those who do to indicate that the bomb has been made more effective. . . .

In the first two years of the atomic era another phenomenon is to be noted. The public, having been warned of the horrible nature of atomic warfare, has done nothing about it, and to a large extent has dismissed the warning from its consciousness. . . .

Americans may be convinced of their determination not to launch an aggressive or preventive war. So they may believe it is superfluous to announce publicly that they will not a second time be the first to use the atomic bomb. But this country has been solemnly invited to renounce the use of the bomb – that is, to outlaw it – and has declined to do so unless its terms* for supranational control are accepted. . . .

In refusing to outlaw the bomb while having the monopoly of it, this country suffers in another respect, in that it fails to return publicly to the ethical standards of warfare formally accepted previous to the last war. . . .

I am not saying that the United States should not manufacture and stockpile the bomb, for I believe that it must do so; it must be able to deter another nation from making an atomic bomb when it also has the bomb. But deterrence should be the only purpose of the stockpile of bombs. . . .

At present the Russians have no evidence to convince them that the American people are not contentedly supporting a policy of military preparedness which they regard as a policy of deliberate intimidation. If they had evidence of a passionate desire by Americans to preserve peace in the one way it can be maintained, by a supranational regime of law, this would upset Russian calculations about the peril to Russian security in current trends of American thought. Not until a genuine, convincing offer is made to the Soviet Union, backed by an aroused American public, will one be entitled to say what the Russian response would be. . . .

It may affront the military-minded person to suggest a regime that does not maintain any military secrets. He has been taught to believe that secrets thus divulged would enable a war-minded nation to seek to conquer the earth. (As to the so-called secret of the atomic bomb, I am assuming the Russians will have this through their own efforts within a short time.) I grant there is a risk in not maintaining military secrets. If a sufficient number of nations have pooled their strength they can take this risk, for their security will be greatly increased. . . .

The atomic scientists, I think, have become convinced that they cannot arouse the American people to the truths of the atomic era by

* This refers to the U.S. plan presented on 14 June 1946 to the U.N. by Bernard Baruch. This proposal never got anywhere.

> logic alone. There must be added that deep power of emotion which is a basic ingredient of religion. It is to be hoped that not only the churches but the schools, the colleges, and the leading organs of opinion will acquit themselves well of their unique responsibility in this regard.

1948.

> We scientists, whose tragic destiny it has been to help make the methods of annihilation ever more gruesome and more effective, must consider it our solemn and transcendent duty to do all in our power in preventing these weapons from being used for the brutal purpose for which they were invented. What task could possibly be more important to us? What social aim could be closer to our hearts?[373]

There follows a pause in Einstein's public engagement in atomic issues until after 31 January 1950, the day on which Truman announced his decision to accelerate the development of a thermonuclear weapon, better known as a hydrogen bomb. Einstein comments: 'General annihilation beckons.' The armaments race had assumed 'a hysterical character. On both sides the means to mass destruction are perfected with feverish haste – behind the respective walls of secrecy.'[374] (He lived to hear of the first successful hydrogen bomb test, on 1 March 1953.)

Reactions to Einstein's statement varied. A Philadelphia paper editorialized: 'Dr. Einstein's vivid warning of atomic dangers ahead falls short of telling us a sure way of avoiding them.'[375]

A representative from Mississippi 'had printed in the Congressional Record excerpts from a report which listed alleged Communist front groups with which Einstein was said to be associated.' A statement attached to the report said in part: 'One of the greatest fakers the world ever knew is Albert Einstein, who should have been deported for his Communist activities years ago. He has been engaged in Communistic activities in this country for a long time. . . . The bunk that he is now spreading is simply carrying out the Communist line.'[376]

This completes what I know about press comments regarding Einstein and the atom, and provides a natural transition to my next subject.

18

The final decade. Einstein on Civil Liberties

It may be said that the Cold War started before the end in 1945 of the Hot War. It is perhaps reasonable to consider the day in 1946 when Churchill gave his 'Iron-Curtain-is-coming-down' speech in Fulton, Missouri, as the beginning of general public consciousness that relations between the West and the Soviet Union were severely strained.

Within the United States, the decade that followed, Einstein's last years, was dominated by anti-communist hysteria, fueled notably by China turning communist, and by Russia exploding its first atomic bomb, both events occurring in 1949.

The political atmosphere in America had already deteriorated earlier, however, largely because of the role of the House Unamerican Activities Committee, set up in 1938 (as an anti-Roosevelt gesture), a forum for right-wingers. Truman's executive order 9835 of March 1947, authorizing the Attorney General to prepare a list of subversive organizations, so-called, had become an added incentive for that Committee's pursuits. In 1950, Joseph ('Joe') McCarthy began his brutal, unscrupulous, and cunning hunt for communists within the US government. This period, one of the blackest in American history, may perhaps be said to have ended on 2 December 1954, when the US Senate condemned its junior member from Wisconsin for conduct 'contrary to Senate traditions' – which is not to say that from then on American anti-communist sentiment abated.

From personal discussions I have vivid recollections of Einstein's moral outrage brought about by this situation. Not least was he indignant about the Government's attitude toward scientists from abroad who wanted to visit the USA but were prohibited from doing so because of alleged communist leanings. Thus he was one of the '34 of the world's leading scientists [who] today directed a concerted and vigorous attack at the visa and passport policies of the United States Government'.[377] According to the press, 'The State Department began [two days later] a thorough study of charges by leading atomic scientists that the Government's visa and passport policies endangered liberty.'[378]

Einstein also gave public expression of his views on the infamous Rosenberg case. Julius Rosenberg and his wife Ethel had been accused of spying for the Soviets. They were brought to trial on 6 March 1951. The next month they were sentenced to death for being guilty under the Espionage Act of 1917 of transmitting secret information on the atomic bomb to the Russians, and immediately imprisoned in Sing Sing. In January 1953 Einstein wrote to Truman: 'Dear Mr. President, My conscience compels me to urge you to commute the death sentence of Julius and Ethel Rosenberg. This appeal to you is prompted by the same reasons which were set forth so convincingly by my distinguished colleague Harold C. Urey in his letter of January 5, 1953, to the *NYT*. Respectfully yours . . .'[379] Urey had strongly recommended a reconsideration of this sentence because, he contended, the evidence was contradictory and inconclusive.

On 19 June 1953 the Rosenbergs died in the electric chair . . .

No action by Einstein during the post-war period created as much sensation and public interest as his letter of 1953 to William Frauenglass, an English teacher at the James Madison High School in Brooklyn, who faced dismissal because of his refusal to testify before the Senate Internal Security subcommittee (the so-called Jenner committee). He had been called in because of lectures he had given six years earlier at an in-service course for teachers, arranged by the Board of Education, on 'Techniques of Intercultural Teaching'. Frauenglass replied by letter: 'On principled constitutional grounds I refuse to answer questions as to political affiliations.' He now faced dismissal under section 903 of the City Charter, which vacates positions of City employees who take the Fifth Amendment.

Frauenglass next approached Einstein, who on 16 May 1953 sent him a letter which was published in full in the press a month later.[380] Here is most of it:

> The problem with which the intellectuals of this country are confronted is very serious. Reactionary politicians have managed to instill suspicion of all intellectual efforts into the public by dangling before their eyes a danger from without. Having succeeded so far, they are now proceeding to suppress the freedom of teaching and to deprive of their positions all those who do not prove submissive, i.e., to starve them out.
>
> What ought the minority of intellectuals to do against this evil? Frankly, I can only see the revolutionary way of non-cooperation in the sense of Gandhi's. Every intellectual who is called before one of the committees ought to refuse to testify, i.e., he must be prepared for jail and economic ruin, in short, for the sacrifice of his personal welfare in the interest of the cultural welfare of his country.
>
> However, this refusal to testify must not be based on the well-known subterfuge of invoking the Fifth Amendment against possible self-incrimination, but on the assertion that it is shameful for a blameless citizen to submit to such an inquisition and that this kind of inquisition violates the spirit of the Constitution.
>
> If enough people are ready to take this grave step they will be successful. If not, then the intellectuals of this country deserve nothing better than the slavery which is intended for them.
>
> P.S. This letter need not be considered 'confidential'.

This letter provoked an explosion of public opinion. Letters *pro* and *con* appeared in numerous newspapers.[381] Senator McCarthy commented that 'anyone who gives advice like Einstein's to Frauenglass is himself an enemy of America.'[382] A week later he said in an interview: 'There is nothing new about that advice. . . . I may say that any American, I don't care whether their name is Einstein or John Jones, who would advise Americans to keep secret information they have about espionage and sabotage – that man is just a disloyal American.'[383] The American Committee for Cultural Freedom called the letter 'ill-considered and irresponsible.'[384] But Bertrand Russell wrote this letter to the *New York Times*: 'Do you condemn the Christian martyrs who refused to sacrifice to the Emperor? . . . I am compelled to suppose that you condemn George Washington . . . As a loyal Briton I of course applaud this view but I fear it may not win much support in your country.'[385] A few months later Irving Adler, another high-school teacher, and Norman London, yet another one, refused to answer questions concerning alleged membership in the Communist Party; both quoted from Einstein's letter.[386]

Two samples of further comments: First an editorial from *The Washington Post*:[387]

> Einstein's Blind Spot
>
> Millions of citizens all over the country share Dr. Albert Einstein's indignation over the extremes to which various congressional inquisitions have been carried. We deplore with him the use of the investigative power to suppress freedom, to invade the responsibilities of the universities and 'to instill suspicion of all intellectual effort'. But we think the great physicist went sadly askew in advising 'intellectuals' who are called before congressional committees to refuse to testify.
>
> In the first place, why should 'intellectuals' have any special immunity that does not apply to other citizens? If we are going to have equality of rights, we must likewise have equality of obligations. And one of the unquestioned obligations of citizenship is to supply information regarding any menace to the national security when so requested by any official body. Suppose that citizens generally followed Dr. Einstein's advice and went to jail in preference to testifying before congressional committees! Our representative system would be paralyzed.
>
> Probably Dr. Einstein did not intend to say that all congressional hearings should be boycotted. But even if we assume that he was talking only about biased inquisitions, could the individual citizen be trusted to decide which committee would get his testimony and which would not? Obviously, if we are going to have orderly government, each summoned witness must be required to speak out before any legislative group – unless the testimony sought would tend to incriminate him. Other means will have to be found to curb the wild men who abuse the investigative power.
>
> Dr. Einstein may not realize it, but he has put himself in the extremist category by his irresponsible suggestion. He has proved once again that genius in science is no guarantee of sagacity in political affairs.

Secondly, an item from a Detroit paper.[388]

> Ex-Senator Hits Einstein
>
> Philadelphia, June 15 – (AP) – Dr. Albert Einstein's advice to American teachers to refuse to testify before a Senate committee

checking into possible Communist affiliations was criticized yesterday as 'an insult' and 'unwarranted interference.'

Former Senator Herbert E. O'Conor (D-Md.) bitterly assailed the noted scientist in a speech before the Pennsylvania Railroad Holy Name Society here.

O'Conor called Einstein's advice one of three 'disturbing developments' which warrant the attention of Americans. The other two, he said, were the 'effort of British trade representatives to increase their dealings with Communist China', and the announcement from Burma that 'it will assume leadership in seeking the admittance of Communist China to the United Nations.'

O'Conor, former head of the Senate Crime Committee, called Dr. Einstein's action an 'indefensible incident'.

This gratuitous counsel comes with bad grace from Dr. Einstein, one who should not be permitted to impede the efforts of officials of our nation to uproot any subversive activities, if they exist, in the universities and colleges of the United States, O'Conor declared.

It is an insult to a senate committee for anyone to defy their legitimate inquiries into the existence of a world-wide conspiracy.

1954. March. A statement by Einstein:

> In principle everybody is equally involved in defending constitutional rights. The intellectuals in the widest sense of the word are, however, in a special position since they have, thanks to their special training, a particularly strong influence on the formation of public opinion. This is the reason why those who are about to lead us toward an authoritarian government are particularly concerned with intimidation and muzzling that group. It is therefore in this situation especially important for the intellectuals to do their duty. I see this duty in refusing to cooperate in any undertaking that violates the constitutional rights of the individual . . . The strength of the Constitution lies entirely in the determination of each citizen to defend it.[389]

April. The California Republication State committee's official organ describes Einstein as a refugee 'free-loader' who considered himself 'above the laws of nations' and to whom 'science is a God' . . . 'This man, who has received from the United States far more than he has given, arrogates to himself the right to instruct his fellows concerning the extent of their obligation to this nation.'[390]

August. Both Houses of Congress approve a modified bill to outlaw the Communist party. Einstein declared, 'It is nonsense', and declined further comment.[391]

November. Another Einstein comment that received wide publicity:

> You have asked me what I thought about your articles concerning the situation of scientists in America. Instead of trying to analyze the problem, I should like to express my feeling in a short remark: If I were a young man again and had to decide how to make a living, I would not

> try to become a scientist or scholar or teacher. I would rather choose to be a plumber or peddler, in the hope of finding that modest degree of independence still available under present circumstances.[392]

The final item of this chapter deals with the only time that I became personally involved in Einstein's relations with the press. It deals with another moment of shame in the American history of civil liberties.

On the morning of Sunday, 11 April 1954, a column appeared in the *New York Herald Tribune*, entitled 'Next McCarthy target: the leading physicists'. I knew what that meant: the Oppenheimer case was about to break.

In December 1953, Robert Oppenheimer had been barred from further work for the United States Atomic Energy Commission, and had been denied further access to classified documents, because allegedly he was a 'poor security risk'. That news had not been made public at that time. I was aware of it, however, and also that Oppenheimer planned to challenge these allegations. That confrontation became known as the Oppenheimer case.

On the evening of that 11 April, I was working in my office at the Institute for Advanced Study in Princeton when the phone rang and a Washington operator asked to speak to Dr Oppenheimer. I replied that Oppenheimer was out of town. (In fact, he was in Washington.) Next the operator asked for Dr Einstein. I told her that Einstein was not at the office and that his home number was unlisted. The operator told me next that her party wished to speak to me. The director of the Washington Bureau of the Associated Press came on the line and told me that the Oppenheimer case would be all over the papers on Tuesday morning. He was eager for a statement by Einstein as soon as possible. I realized that pandemonium on Mercer Street the next morning might be avoided by a brief statement that evening and therefore said that I would talk it over with Einstein and would call back in any event. I drove to Mercer Street and rang the bell; Helen Dukas let me in. I apologized for appearing at such a late hour and said it would be good if I could talk briefly with the professor, who meanwhile had appeared at the top of the stairs dressed in his bathrobe and asked, '*Was ist los?*' – 'What is going on?' He came down and so did his stepdaughter Margot. After I told him the reason for my call, Einstein burst out laughing. I was somewhat taken aback and asked him what was so funny. He said that the problem was simple. All Oppenheimer needed to do, he said, was go to Washington, tell the officials that they were fools, and then go home. On further discussion, we decided that a brief statement was called for. It said: 'I can only say I have the greatest respect and warmest feelings for Dr. Oppenheimer. I admire him not only as a scientist but also as a man of great human qualities.'[392a] We drew it up, and Einstein read it over the phone to the AP director in Washington.

The next day Helen Dukas was preparing lunch when she saw cars in front of the house and cameras being unloaded. In her apron (she told me) she ran out of the house to warn Einstein, who was on his way home. When he arrived at the front door, he declined to talk to reporters.

The May 1954 issue of the *Bulletin of the Atomic Scientists* included statements on the case by a number of outstanding scientists. Einstein's contribution, terse and brief, provides a fitting conclusion to this sad chapter. It reads, in full:

> The systematic and widespread attempt to destroy mutual trust and confidence constitutes the severest possible blow against society.[393]

19

Einstein and the Jews Death of Einstein

Not since Chapter 9 have I taken note of Einstein's pronouncements on Jewish issues. There I had recalled that in his early years Einstein had already been aware of being a Jew and of the existence of anti-Semitism but had not paid particular attention to his descent; that his first exposure to Zionism dated from the years just after the First World War, and that he never joined any Zionist organization. I have also given a brief account of his one and only visit to Palestine in 1923; and of his first comments to the press, dating from 1925, regarding Jewish questions. I shall now follow his later public comments until, literally, the end of his life.*

1926. From 1921 onwards, Arabs in Palestine frequently engaged in anti-Jewish riots, some of them bloody. This was one of the reasons for occasional proposals to colonize Jews elsewhere. In 1926 Einstein commented as follows on the idea to do so in Russia:

> Although I believe that it is only in Palestine that work of lasting value can be achieved and that everything that is done in the Diaspora countries is only a palliative, I nevertheless hold that the efforts which are being made to colonize Jews in Russia must not be opposed because they aim at assisting thousands of Jews whom Palestine cannot immediately absorb. On this ground, these efforts seem to me worthy of support. I do not, therefore, believe that the money which is expended in Russia on Jewish colonization is being wasted. Whether the necessary guarantees exist for the success of this colonization work, I cannot say without first having been on the spot. But if the colonization is successful, it will ultimately be of benefit also to us because it will mean a strengthening of the Jewish people and every effort, every

* See also the collection of Einstein's essays *On Zionism*.[107]

factor, which strengthens our people, even if only morally or indirectly, is justified.[394]

1929. In 1929 there occurred a series of extremely violent attacks by Arabs against Jewish settlers. To add insult to injury, certain British circles used the occasion for making anti-Zionist statements. Einstein reacted with an indignant letter to a British newspaper which included these lines:

> Arab mobs, organised and fanaticised by political intriguers working on the religious fury of the ignorant, attacked scattered Jewish settlements and murdered and plundered wherever no resistance was offered. In Hebron, the inmates of a rabbinical college, innocent youths who had never handled weapons in their lives, were butchered in cold blood; in Safed the same fate befell aged rabbis and their wives and children. Recently some Arabs raided a Jewish orphan settlement where the pathetic remnants of the great Russian pogroms had found a haven of refuge. Is it not then amazing that an orgy of such primitive brutality upon a peaceful population has been utilised by a certain section of the British press for a campaign of propaganda directed, not against the authors and instigators of these brutalities, but against their victims?[395]

1930. In January Einstein addressed the Arabs directly:

> One who, like myself, has cherished for many years the conviction that the humanity of the future must be built up on an intimate community of the nations, and that aggressive nationalism must be conquered, can see a future for Palestine only on the basis of peaceful co-operation between the two peoples who are at home in the country. For this reason I should have expected that the great Arab people will show a truer appreciation of the need which the Jews feel to re-build their national home in the ancient seat of Judaism; I should have expected that by common effort ways and means would be found to render possible an extensive Jewish settlement in the country. I am convinced that the devotion of the Jewish people to Palestine will benefit all the inhabitants of the country, not only materially, but also culturally and nationally. I believe that the Arab renaissance in the vast expanse of territory now occupied by the Arabs stands only to gain from Jewish sympathy. I should welcome the creation of an opportunity for absolutely free and frank discussion of these possibilities, for I believe that the two great Semitic peoples, each of which has in its way contributed something of lasting value to the civilisation of the West, may have a great future in common, and that instead of facing each other with barren enmity and mutual distrust, they should support each other's national and cultural endeavors, and should seek the possibility of sympathetic cooperation. I think that those who are not actively engaged in politics should above all contribute to the creation of this atmosphere of confidence.
>
> I deplore the tragic events of last August not only because they revealed human nature in its lowest aspects, but also because they have

estranged the two peoples and have made it temporarily more difficult for them to approach one another. But come together they must, in spite of all.[396]

September. In an address to the first international congress of Palestine workers, held in Berlin: 'It does not matter how many Jews are in Palestine but it does matter what they produce there. That should be something the Jews of the whole world can point to as ideal creative work and with which they can identify themselves.'[397]

October. In a speech delivered on 29 October at the Savoy Hotel in London, Einstein said in part:

> The position of our scattered Jewish community is a moral barometer for the political world. For what surer index of political morality and respect for justice can there be than the attitude of the nations toward a defenceless minority, whose peculiarity lies in their preservation of an ancient cultural tradition?
>
> This barometer is low at the present moment, as we are painfully aware from the way we are treated. But it is this very lowness that confirms me in the conviction that it is our duty to preserve and consolidate our community. Embedded in the tradition of the Jewish people there is a love of justice and reason which must continue to work for the good of all nations now and in the future . . .
>
> Remember that difficulties and obstacles are a valuable source of health and strength to any society. We should not have survived for thousands of years as a community if our bed had been of roses; of that I am quite sure.[398]

1934. April. Einstein gave his opinion on revisionism, a right-wing Jewish movement that advocated and executed terrorist actions aiming at the realization of a Jewish state in Palestine.

> Revisionism is the modern embodiment of those harmful forces which Moses with foresight sought to banish when he formulated his model codes of social law.
>
> The secret of our apparently inexhaustible vitality lies in our strong traditions of social justice and of modest service to our immediate community and society as a whole.
>
> The Jews must beware of viewing Palestine merely as a place of refuge.[399]

Also in 1934 Einstein issued a statement entitled 'Let's not forget':

> If we as Jews can learn anything from these politically sad times, it is the fact that destiny has bound us together, a fact which, in times of quiet and security, we often so easily and gladly forget. We are accustomed to lay too much emphasis on the differences that divide the Jews of different lands and different religious views. And we forget often that it is the concern of every Jew, when anywhere the Jew is hated and treated unjustly, when politicians with flexible consciences set into

> motion against us the old prejudices, originally religious, in order to concoct political schemes at our expense. It concerns every one of us because such diseases and psychotic disturbances of the folk-soul are not stopped by oceans and national borders, but act precisely like economic crises and epidemics.[400]

1935. On 24 March Einstein spoke at a Purim dinner of the German-Jewish club in New York: 'There are no German Jews, there are no Russian Jews, there are no American Jews. Their only difference is their daily language. There are in fact only Jews.'[401]

April. At a Passover celebration in New York, Einstein again condemned revisionism and urged Jewish–Arab unity. He reminded the audience that the founders of the Zionist movement worked for the traditional ideals of justice and the selfless love of mankind.[402]

June.

> The intellectual decline brought on by shallow materialism is a far greater menace to the survival of the Jew than the numerous external foes who threaten its existence with violence. We must never forget that through all the severe afflictions for twenty centuries our ancestors found consolation, refuge and strength in the fostering of our spiritual traditions.[403]

1938. January. Einstein sent greetings to the National Council for Jewish Women, in session in Pittsburgh:

> Mutual assistance is, God be thanked, our one weapon in our bitter struggle for existence. Weakened through dispersion in countless factions, we, nonetheless, remain united through the fairest of all duties – the duty of unselfish mutual aid. Never has Jewry denied itself to the demands of this duty.[404]

April. Einstein spoke in German at the Seder by the National Labor Committee for Palestine at the Hotel Astor in New York.

> To be a Jew . . . means first of all, to acknowledge and follow in practice those fundamentals in humaneness laid down in the Bible – fundamentals without which no sound and happy community of men can exist.
>
> We meet today because of our concern for the development of Palestine. In this hour one thing, above all, must be emphasized: Judaism owes a great debt of gratitude to Zionism. The Zionist movement has revived among Jews the sense of community . . .
>
> Now the fateful disease of our time – exaggerated nationalism, borne up by blind hatred – has brought our work in Palestine to a most difficult stage. Fields cultivated by day must have armed protection at night against fanatical Arab outlaws . . .
>
> Just one more personal word on the question of partition. I should much rather see reasonable agreement with the Arabs on the basis of

living together in peace than the creation of a Jewish state . . . We are no longer the Jews of the Maccabee period. A return to a nation in the political sense of the word would be equivalent to turning away from the spiritualization of our community which we owe to the genius of our prophets. If external necessity should after all compel us to assume this burden, let us bear it with tact and patience.[405]

Also, a message by Einstein urging Great Britain not to yield to the pressure of terrorism:

> Our bitter appeal is addressed to the nations and to England, and our demand should have the support of unwritten law and justice. We ask England not to compel by the sword what she has promised, but not to permit a minority to impose its will through terror on the majority of Arabs and Jews.[406]

November. 'Why do they hate the Jews?', a long magazine article by Einstein, was published in a New York magazine.[407] From its text:

> Why did the Jews so often happen to draw the hatred of the masses? Primarily because there are Jews among almost all nations and because they are everywhere too thinly scattered to defend themselves against violent attack.
>
> A few examples from the recent past will prove the point: Toward the end of the nineteenth century the Russian people were chafing under the tyranny of their government. Stupid blunders in foreign policy further strained their temper until it reached the breaking point. In this extremity the rulers of Russia sought to divert unrest by inciting the masses to hatred and violence toward the Jews. These tactics were repeated after the Russian government had drowned the dangerous revolution of 1905 in blood – and this maneuver may well have helped to keep the hated regime in power until near the end of the World War.
>
> When the Germans had lost the World War hatched by their ruling class, immediate attempts were made to blame the Jews, first for instigating the war and then for losing it. In the course of time, success attended these efforts . . .
>
> The crimes with which the Jews have been charged in the course of history – crimes which were to justify the atrocities perpetrated against them – have changed in rapid succession. They were supposed to have poisoned wells. They were said to have murdered children for ritual purposes. They were falsely charged with a systematic attempt at the economic domination and exploitation of all mankind. Pseudo-scientific books were written to brand them an inferior, dangerous race. They were reputed to foment wars and revolutions for their own selfish purposes. They were presented at once as dangerous innovators and as enemies of true progress. They were charged with falsifying the culture of nations by penetrating the national life under the guise of becoming assimilated. In the same breath they were accused of being so stubbornly

inflexible that it was impossible for them to fit into any society . . .

[What is the basis for these allegations?]

The members of any group existing in a nation are more closely bound to one another than they are to the remaining population. Hence a nation will never be free of friction while such groups continue to be distinguishable. In my belief, uniformity in a population would not be desirable, even if it were attainable. Common convictions and aims, similar interests, will in every society produce groups that, in a certain sense, act as units. There will always be friction between such groups – the same sort of aversion and rivalry that exists between individuals . . .

Were anyone to form a picture of the Jews solely from the utterances of their enemies, he would have to reach the conclusion that they represent a world power. At first sight that seems downright absurd; and yet, in my view, there is a certain meaning behind it. The Jews as a group may be powerless, but the sum of the achievements of their individual members is everywhere considerable and telling, even though these achievements were made in the face of obstacles. The forces dormant in the individual are mobilized, and the individual himself is stimulated to self-sacrificing effort, by the spirit that is alive in the group.

Hence the hatred of the Jews by those who have reason to shun popular enlightenment. More than anything else in the world, they fear the influence of men of intellectual independence . . .

1939. March. From a radio address for United Jewish Appeal, broadcast on the 22nd:

In the past we were persecuted *despite* the fact that we were the people of the Bible; today, however, it is just *because* we are the people of the Book that we are persecuted. The aim is to exterminate not only ourselves but to destroy, together with us, that spirit expressed in the Bible and in Christianity which made possible the rise of civilization in Central and Northern Europe. If this aim is achieved, Europe will become a barren waste. For human community life cannot long endure on a basis of crude force, brutality, terror, and hate . . .

One of the most tragic aspects of the oppression of Jews and other groups has been the creation of a refugee class. Many distinguished men in science, art, and literature have been driven from the lands which they enriched with their talents . . . As one of the former citizens of Germany who have been fortunate enough to leave that country, I know I can speak for my fellow refugees, both here and in other countries, when I give thanks to the democracies of the world for the splendid manner in which they have received us. We, all of us, owe a debt of gratitude to our new countries, and each and every one of us is doing the utmost to show our gratitude by the quality of our contributions to the economic, social, and cultural work of the countries in which we reside.[408]

May. A radio address broadcast from Einstein's home to a meeting of the Jewish National Workers Alliance, held in Town Hall, New York:

'Remember in the midst of your justified embitterment that England's opponents are also our bitterest enemies and that, in spite of everything, the maintenance of England's position is of utmost importance to us.'[409]

(On 5 April 1920, at the San Remo Conference, Britain had been assigned by the League of Nations as mandatory power for Palestine, according to principles laid down in the Treaty of Versailles. In 1939 Britain was still in control over that area.)

On 3 September the Second World War broke out.

1940. At a testimonial dinner to Einstein, given by the friends of the Haifa Technion (an Institute of Technology), Einstein said:

> I can remember very well the time when Jews in Germany laughed over Palestine. I remember, when I spoke with Rathenau about Palestine, he said: 'Why go to this land that is only sand and worth nothing and which can never be developed?' This was his idea. But, if he had not been murdered, he probably would now be in Palestine. You can therefore see that the development of Palestine is of real tremendous importance for all of Jewry.[410]*

1944. June. Message by Einstein to a dinner by the American Fund for Palestinian Institutions:

> The spirit of the Jews in Palestine has remained fresh and resilient. I have no doubt that they will succeed in a good measure of cooperation with the Arab people if only both our people and the Arabs succeed in conquering that childhood complaint of a narrow-minded nationalism imported from Europe and aggravated by professional politicians. Both peoples, it is to be hoped, will soon recognize that no rigid legal formula but only a lively mutual understanding and faithful cooperation in the daily tasks can open the right way.[411]

1946. January. Papers carry a detailed account[412] of Einstein's testimony, given in Washington, D.C., before the Anglo-American Committee of Inquiry on Palestine.

'Smiling benignly, Professor Einstein launched into the most wholehearted denunciation of British colonial policy that the committee has yet heard.' He makes four points.

> (1) He is convinced, though regretfully, that British colonial policy is such as to render Great Britain unfit for further administration of her mandate over Palestine. (2) A trusteeship should be set up by the United Nations to administer Palestine and it should not be confined to any single power, including the United States. (3) The great majority

* On 24 June 1922 Walther Rathenau, the German Foreign Minister, a Jew, and an acquaintance of Einstein, had been assassinated in Berlin.

> of Jewish refugees in Europe should be settled in Palestine. (4) He has never seen any necessity for a Jewish commonwealth such as is advocated by the Zionist Organization for Palestine. He 'repeated two or three times his belief that inquiry commissions such as the one before which he was appearing have been created only to give the impression of goodwill without any intention on the part of responsible authorities to pay any attention to their findings and recommendations.' Richard Grossman, MP, got the witness to agree that it was not 'a British imperialist fiction' that Arabs might shoot Jewish refugees if they came in large numbers. When asked what would happen if the Arabs resisted Jewish immigration, Einstein replied 'They won't if they are not instigated. If people work together, they won't worry who has the larger number.' Also 'the State [of Israel] idea is not according to my heart. I cannot understand why it is needed. It is connected with narrow-mindedness and economic obstacles. I believe it is bad. I have always been against it . . . [It is] an imitation of Europe – the end of Europe was brought about by nationalism.'

February. In a letter to the Progressive Palestine Association, Einstein expressed the belief that 'A Government in Palestine under the United Nation's direct control and a constitution assuring Jews' and Arabs' security against being outvoted by each other would solve the Jewish–Arab difficulties.'[413]

1947. Einstein sends a message to a dinner at the Waldorf Astoria Hotel in New York in honor of Weizmann: 'In these days of fateful decision you have presented our case before the world with a vision that no one among us could muster.'[414]

On 29 November 1947, four days after that dinner, the United Nations General Assembly passed a resolution to the effect that the British Mandate for Palestine should be terminated. This decision was the result of a recommendation by a special UN Committee for Palestine which had started deliberations the preceding May. Arab violence broke out immediately after the news of the resolution became known.

1948. The British Mandate ended on 15 May 1948. On 14 May the State of Israel was officially proclaimed. On the 15th, Israel was attacked by combined forces from Egypt, Transjordan, Syria, Lebanon, and Iraq. The resulting War of Independence ended formally in July 1949, in defeat of the Arab Alliance.

A month before the outbreak of that War, Einstein had sent the following letter to the *New York Times*:

> Both Arab and Jewish extremists are today recklessly pushing Palestine into a futile war . . . We feel it to be our duty to declare emphatically that we do not condone methods of terrorism and of fanatical national-

> ism any more if practiced by Jews than if practiced by Arabs . . . A decisive victory by either would yield a corroding bitterness . . . We appeal to the Jews in this country and in Palestine not to permit themselves to be driven into a mood of despair or false heroism which eventually results in suicidal measures.[415]

1949. March. Upon receiving an honorary degree from the Hebrew University, Einstein declared: 'The wisdom and moderation the leaders of the new State have shown gives me confidence that gradually relations will be established with the Arab people which are based on fruitful cooperation and mutual trust.'[416]

November. Einstein was re-elected Chairman of the National Council of Friends of the Hebrew University.[417] On the 27th of that month, Einstein made a radio broadcast for the United Jewish Appeal in which he said in part:

> There is no problem of such overwhelming importance to us Jews as consolidating that which has been accomplished in Israel with amazing energy and an unequaled willingness for sacrifice. May the joy and admiration that fill us when we think of all that this small group of energetic and thoughtful people has achieved give us the strength to accept the great responsibility which the present situation has placed upon us. [They have created] a community which conforms as closely as possible to the ethical ideals of our people as they have been formed in the course of a long history.
>
> One of these ideals is peace, based on understanding and self-restraint, and not on violence. If we are imbued with this ideal, our joy becomes somewhat mingled with sadness, because our relations with the Arabs are far from this ideal at the present time. It may well be that we would have reached this ideal had we been permitted to work out, undisturbed by others, our relations with our neighbors, for we *want* peace and we realize that our future development depends on peace.
>
> It was much less our own fault or that of our neighbors than of the Mandatory Power that we did not achieve an undivided Palestine in which Jews and Arabs would live as equals, free, in peace. If one nation dominates the other nations, as was the case in the British Mandate over Palestine, she can hardly avoid following the notorious device of *Divide et Impera*. In plain language this means; create discord among the governed people so they will not unite in order to shake off the yoke imposed upon them. Well, the yoke has been removed, but the seed of dissension has borne fruit and may still do harm for some time to come – let us hope not for too long.
>
> The economic means of the Jewish Community in Israel do not suffice to bring this tremendous enterprise to a successful end. For a hundred thousand out of more than three hundred thousand persons who immigrated to Israel since May, 1948, no homes or work could be made available. They had to be concentrated in improvised camps under conditions which are a disgrace to all of us.
>
> It must not happen that this magnificent work breaks down because

Einstein with Ben Gurion in the backyard of the Mercer Street home, May 1951. (© Mrs Alan Richards. Photo courtesy of the Archives, Institute of Advanced Study, Princeton.)

> the Jews of this country do not help sufficiently or quickly enough. Here, to my mind, is a precious gift with which all Jews have been presented: the opportunity to take an active part in this wonderful task.[418]

1951. Einstein bought the 200,000th $500 State of Israel bond of a recent issue.[419]

1953. In a plea of support for the Hebrew University, Einstein said: 'For the young State to achieve real independence, and conserve it, there must be a group of intellectuals and experts produced in the country itself.'[420]

1954. Einstein spoke at a planning conference in Princeton of the American Friends of the Hebrew University.

> Israel is the only place on earth where Jews have the possibility to shape public life according to their own traditional ideals . . . In our tradition it is neither the ruler nor the politician, neither the soldier nor the merchant, who represents the ideal. The ideal is represented by the teacher who . . . enrich[es] the intellectual, moral, and artistic life of the people. This implies a definite repudiation of what is commonly called 'materialism'. Human beings can attain a worthy and harmonious life only if they are able to rid themselves, within the limits of human

> nature, of the striving for the wish fulfillments of material kinds. The goal is to raise the spiritual values of society.[421]

1955. On the morning of Wednesday 13 April 1955, the Israeli consul called on Einstein at his home in order to discuss the draft of a statement Einstein intended to make on television and radio on the occasion of the forthcoming anniversary of Israel's independence. The incomplete draft ends as follows.

> No statesman in a position of responsibility has dared to take the only promising course [toward a stable peace] of supranational security, since this would surely mean his political death. For the political passions, aroused everywhere, demand their victims.[422]

These may well be the last phrases Einstein committed to paper. Two hours after this visit, Einstein was fatally stricken. He died in the early morning hours of 18 April. In the afternoon of that day he was cremated. The ashes were scattered in an unknown place.

20

To recapitulate

As has been stated in the Introduction, the purpose of this essay is to demonstrate that Einstein is a creation of the media insofar as he is a public figure. Up to this point, 297 references in newspapers and magazines have been used to document that.

We have now followed Einstein up to his death. The final chapter, to follow, will illustrate that the media have perpetuated the Einstein legend into the present, and will no doubt continue to do so for a long time to come.

I shall try next to summarize what we have learned from Einstein's numerous appearances in the press.

To begin with, the coverage of Einstein's science, the cause for his world fame (Chapters 3 and 4). His science is a topic for which reporters remained hungry. At the same time that the quality of his scientific contributions waned, the press grew ever more ecstatic about this work (Chapters 12, 13). Reporting on science has never been one of the media's fortes.

On the political scene, Einstein's pronouncements at first centered on pacifism. His initial position, formulated shortly after the First World War, was that in a world tired of war man should strive for improved understanding and cooperation between nations. Toward the late 1920s he became much more radical, however, advocating refusal by the individual to bear arms (Chapters 5, 11). The rise of the Nazis made him change his mind, but after the Second World War he returned to his earlier outspoken pacifist views (Chapter 17).

Einstein's next main political theme was supranationalism. He came out in favor of establishing a world organization that would not only arbitrate international conflicts but furthermore, if necessary, would be in a position to implement its decisions by the force of armies that would be available to and commanded by that world organization (Chapters 11, 17). No such military option was available to the League of Nations, an organization of which Einstein held dim views (Chapter 11).

The third and last focus of Einstein's political opinions concerned civil liberties, most acutely during his final years which coincided with the McCarthy era (Chapter 18).

Einstein's utterances on political matters did not always address the immediately practicable, and I do not think that, on the whole, they were very influential. Bear in mind that in science one seeks the answer to a given problem, whereas in politics a problem has no answer – only a compromise; and that to Einstein nothing was more alien than to settle any issue by compromise, in his life or in his science.

This has led to the fairly widespread view that Einstein was politically naïve. The best response to this opinion I can think of is to quote a few lines from Hans Christian Andersen's well-known story *The Emperor's New Clothes*.

> No one would admit he could see nothing, for that would be admitting he was a fool.
> 'But he's got nothing on', marvelled a little child.
> 'Ah, just listen to this innocent one, this naïve one', said its father.
> And one person began to whisper to another: 'He's got nothing on, a little child says he's got nothing on.'
> 'But he has nothing on', cried everyone at last.

Einstein was like that little boy, honest, unpretentious – but not naïve. Time and again, his political opinions have proved to be valid, not in the short term but in the long run. Only in the 1990s have we begun to see a world organization whose political decisions are backed up by military force. And would Oppenheimer not have been far better off had he followed Einstein's advice (Chapter 18) to tell the Washington officials that they were fools – as everybody now knows they indeed were?

Related to, but also transcending, Einstein's political opinions was his attitude to the problems of the Jews. Not until he was in his thirties did the full weight and urgency of ameliorating the lot of the Jews dawn on him. As he wrote in 1929:

> When I came to Germany [in 1914, at age 35] I discovered for the first time that I was a Jew, and I owe this discovery more to Gentiles than to Jews . . .
>
> I saw worthy Jews basely caricatured, and the sight made my heart bleed. I saw how schools, comic papers, and innumerable other forces of the Gentile majority undermined the confidence even of the best of my fellow-Jews, and felt that this could not be allowed to continue.
>
> Then I realized that only a common enterprise dear to the heart of Jews all over the world could restore this people to health. It was a great achievement of Herzl's to have realized and proclaimed at the top of his voice that, the traditional attitude of the Jews being what it was, the establishment of a national home or, more accurately, a center in Palestine, was a suitable object on which to concentrate our efforts.
>
> All this you call nationalism, and there is something in the accusation. But a communal purpose without which we can neither live nor die in this hostile world can always be called by that ugly name. In any case it is a nationalism whose aim is not power but dignity and health. If we did not have to live among intolerant, narrow-minded, and violent people, I should be the first to throw over all nationalism in favor of universal humanity.[423]

As these lines make clear, in Einstein's mind Zionism as he interpreted it and supranationalism were not at variance with each other. Nor did he look upon Judaism as a religious creed. To him, the Jewish tradition was 'a sort of intoxicated joy and amazement at the beauty and grandeur of this world, of which man can form just a faint notion. This joy is the feeling from which true scientific research draws its spiritual sustenance, but which also seems to find expression in the song of birds. To tack this feeling to the idea of God seems mere childish absurdity.'[424]

Thus it seems to me fair to summarize Einstein's views on the Jews and their fate like this. Their common bond is spiritual rather than more narrowly religious. Palestine, later Israel, should be a haven for the persecuted, to be sure, but should above all be the region where Jewish spiritual values are kept alive. On numerous occasions Einstein expressed sympathy with the plight of the Arabs in Israel – though of course not with their terror tactics. He hoped for full cooperation of Jews with Arabs but realized that for the time being that could not come to pass. I am sure he would have been disgusted with the later governments of Begin and Shamir had he lived to see their methods – and would have said so publicly.

The purpose of inserting occasional varia has been to show that Einstein was news practically whatever he did, whether meeting other world figures like Chaplin, Rockefeller, or Bernard Shaw (Chapter 12), or whether he corresponded with Freud (Chapter 13) or President Roosevelt (Chapter 15), whether he opened the 1939 New York World Fair (Chapter 16), took a bath or pronounced on the death penalty (Chapter 11), or whether he helped a schoolgirl or had his brain waves recorded (Chapter 17). And, of course, his opinions on political matters other than those I just summarized, and on education, were always considered worth quoting.

Next to his scientific talents, Einstein's greatest gift was his writing in German, the language with which he was most comfortable throughout his life and which he used almost invariably when committing his thoughts to paper, whether these dealt with science or with other subjects. Many pieces by his hand that appeared in print in English are in fact translations from a first German draft. In the preceding I had of course to render everything in English. I hope this will not have obscured too much Einstein's original graceful and occasionally witty style.

21

And the Show Goes On

Einstein's death did not diminish the attention the press paid to him. In some respects the opposite is true.

1. Post mortems

On 19 April 1955, the day after his death, the *New York Times* reported the news on its front page, including a lovely picture of him. That same issue also contains tributes by the Presidents of the United States and West Germany, and the Prime Ministers of Israel, France, and India, as well as one by his colleagues on the faculty of the Institute for Advanced Study: '[He] has passed from among us into the indelible record of history, where his lofty place has long been assured.' There was also an editorial which included this phrase: 'Mathematical physicists in Einstein's class are the epic poets of our time.'

Later in that same month of April we find numerous press items concerning the hero who had departed in body but not in spirit.

Niels Bohr, the other great leader of twentieth-century physics, said:

> Through Albert Einstein's work the horizon of mankind has been immeasurably widened, at the same time as our world picture has attained [through E.'s work] a unity and harmony never dreamed of before. The background for such achievement was created by preceding generations of the worldwide community of scientists, and its full consequences will only be revealed to coming generations.
>
> The gifts of Einstein are in no way confined to the sphere of science. Indeed his recognition of hitherto unheeded assumptions in even our

> most elementary and accustomed assumptions means to all people a new encouragement in tracing and combating the deep rooted prejudices and complacencies inherent in every national culture.[425]

Prime Minister Nehru of India stated that he had just received a letter from Einstein containing proposals for a five-year truce in the Cold War.[425]

A recollection: Einstein on clothes. 'It would be a sad situation if a bag was better than the meat wrapped in it.'[425]

Another recollection. 'In 1921 when Dr. Einstein made his first visit to the United States, interest in his theory and its meaning was so great that Representative J.J. Kindred of New York requested the Speaker of the House to insert a popular presentation of the relativity theory in the Congressional Record.'[425]

Two days later: 'Einstein's country house in Caputh near (East) Berlin will become a memorial, to be created by East Germany.'[426]

The next day: 'When asked why could man discover atoms but not the means for controlling them, Einstein replied: "That is simple, my friend. It is because politics is more difficult than physics."'[427]

May 1955. A moon crater known as 'Simpelius D' will be renamed Einstein. 'It is near a crater named after Sir Isaac Newton.'[426]

Also in May, David J. Levy, attorney for Einstein's family, states: 'I am requested and authorized to say that the ashes have now been privately, finally, and irrevocably been disposed of in conformity with the wishes of Professor Einstein. Since the where and how are matters of detail, they should not be of public interest, for the ultimate fact is that no physical traces are left anywhere.'[428]

August 1955. 'The first full-scale hydrogen bomb explosion in the Pacific in late 1952 produced two new chemical elements, atomic scientists disclosed today. The researchers recommended that the new elements be named after Dr. Albert Einstein and Dr. Enrico Fermi, who played major roles in the birth of the atomic age.'[429]

It has indeed come to pass that the elements with atomic numbers 99, 100 are now known as einsteinium and fermium respectively.

2. Coming out of the woodwork

In later years press items have appeared that shed additional light on events in Einstein's life. I have collected five such items.

Einstein early on served as honorary chairman of the presidium of the Yiddish Scientific Institute (YIVO). The September 1979 issue of the organization's bulletin contains a copy of a letter by Einstein dated 8 April 1929:

> As is well known, the impoverishment of Eastern Europe has caused an economic depression of the Jewish masses. The most urgent demand was, in first instance, to care for maintaining their naked existence. Man does not live from bread alone, however, especially not the Jew. When therefore a small group of man of the intellectual elite have joined in efforts to maintain and possibly develop further the intellectual and moral traditions of our people, then those attempts deserve sympathy and active help from all who can do so.[430]

From a French journal[431] of 1965, concerning an event that occurred during Einstein's December 1930 visit to New York (Chapter 12):

> A pastor who was in the process of designing a new Protestant church, Riverside Church in Manhattan, had decided to decorate its facade with statues of great men who had honored humanity. On requesting from the best known scholars a list of fourteen names, only one appeared in every answer: Albert Einstein. When the pastor showed him his effigy in stone, he commented: 'I could have imagined that one day one would make me a Jewish saint, but never that I would become a Protestant saint.

A recollection in 1989 of a man who attended the opening ceremony of the New York World Fair's Palestine pavilion fifty years earlier, at which Einstein spoke (Chapter 14). 'Einstein was for us perhaps the most famous Jew in the world. His English address with a German accent was incomprehensible to the close to 100,000 people who had assembled. But the emotion was there nevertheless.' He remembers throwing himself into the crowd to shake Einstein's hand, but the crush was too great.[432]

In 1947 Einstein accepted membership of the advisory committee for the department of applied mathematics of the Weizmann Institute in Rehovoth, Israel. At an early meeting (late 1940s) of the committee Einstein objected to the proposal for an electronic computer at the department. He could not see why poor Palestine should enter into such a venture if one such instrument did not yet exist in the whole of Europe. In the end he was, however, persuaded to agree.[433]

Finally, a letter written in 1953 by Einstein to a man active in promoting vegetarianism. The letter was published in 1957. Its main point follows:

> When you buy a piece of morass for planting your cabbages and apples, then you have first to dehydrate the ground. That will kill the water fauna. Later you will have to kill caterpillars etc. which would eat up your meagre food. If you wish to avoid all that, then you have to commit suicide and you will leave alive those to whom all higher moral principles are unknown and inaccessible.
>
> With all respect for moral precepts one must admit that, regarding those, man's existence needs compromises which it is necessary but not easy to determine. In any event one can not reach those compromises by moral indignation alone.[434]

3. Einstein in the arts; and in advertising

Einstein as inspiration for artistic expression is a subject worthy of a book by itself.* My own comments will be extremely brief, because I am not at all sufficiently familiar with the subject to do it justice, and also, to be frank, because the topic does not move me much. Moreover, it may well be argued that the inclusion of Einstein as he appears in the arts stretches the point of describing him as he is seen by the press. I confine myself to just a few examples from the time after his death.[436]

In *belles lettres*: Lawrence Durrell's acknowledgement of inspiration derived from relativity theory. In the preface to part 2 of his *Alexandria Quartet* he wrote:

> Modern literature offers us no Unities, so I have turned to science and am trying to complete a four-decker novel whose form is based on the relativity proposition.
>
> Three sides of space and one of time constitute the soup-mix recipe of a continuum. The four novels follow this pattern.
>
> The three first parts, however, are to be deployed spatially . . . and are not linked in a serial form. They interlap, interweave, in a purely spatial relation. Time is stayed. The fourth part alone will represent time and be a true sequel.[437]

Relativity recurs in the text: 'The Relativity proposition was directly responsible for abstract painting, atonal music, and formless . . . literature.'[438] And in part 4: 'If I wrote I would try for a multi-dimensional effect in character.'[439] I am very fond of the *Quartet*, in spite of these quotations.

In the theatre: Friedrich Dürrenmatt's play *The physicists*[440] features Newton, Einstein, and the mathematician August Möbius as inmates of an insane asylum even though they are sane. The three men are contending for control of the secrets of nuclear power. The play ends with Einstein playing the violin.

In film: The BBC-TV production *Einstein, the story of the man told by his friends*, released in 1969.

In opera: *Einstein on the Beach* by Philip Glass, first performed in 1976.

Among Einstein's appearances in the arts one may perhaps also include his renderings on numerous postage stamps between the years 1956 and 1976, including those from Argentina, Canada, Chad, the Comoro Islands, Ghana, Israel, Mali, Nicaragua, Paraguay, Poland, and the USA.[441]

One summer day I was driving along a Danish highway when a Swedish tourist bus passed me by. Looking to my left I saw good old Albert looking

* A good step in this direction has already been made; see ref. 435, which has been of great help to me.

at me, his portrait painted not once but three times on the side of the bus. That was an unexpected but not surprising sight.

No medium has done more than advertising for promoting Einstein to a universal icon. All of us have frequently seen him staring at us from the advertisement pages of daily papers, weeklies, magazines, billboards or what have you. Or, rather – I find this disturbing, having his great face firmly in my memory – someone made up to look like him. That was no doubt done in order to avoid requests for permission to reproduce actual photographs.

The products these images aim to promote range far and wide, from women's hosiery to fast computers. The essence of the message is always the same: the buyer will demonstrate intelligence or smart thinking by purchasing this or that product. To give but one among legions of examples: you see a picture showing a monkey on the left, Einstein on the right, both holding a filled beer mug. Caption under the monkey: 'Instinct says beer'; under Einstein: 'Reason says Carlsberg.'[442]

4. The Centennial

Einstein was born in Ulm, Germany, on 14 March 1879, at 11.30 a.m. Accordingly the year 1979 was rich in centennial celebrations, some solemn, some learned, some festive, most a mixture of the above.

Ulm celebrated its native son, Zürich 'the greatest of all its adoptive sons'[443] with a meeting in the Auditorium Maximum of the ETH (the Federal Institute of Technology) and an Einstein exhibition. A newspaper photograph[443] shows Thomas Martin Einstein – Albert's great-grandson, born within months after Einstein's death – attending the opening of that show.

Centennial conferences were held in Berlin, Jerusalem, Princeton, and no doubt elsewhere as well. Many of the highest government authorities participated.

The United States produced a 15-cent Einstein Centennial stamp, showing the fine portrait of Einstein as an old man by the photographer Hermann Landshoff. (Date and place of first issue: 14 March, Princeton.) The People's Republic of China also came out with a stamp, showing a good likeness of the man above his formula $E = mc^2$. Perhaps other countries have done similarly.

On 19 February *Time* magazine came out with yet another Einstein cover story. So did *Look* magazine on 2 April.

The Institute for Advanced Study in Princeton not only organized a scientific Centennial Conference, but also arranged or co-sponsored other events: a 90-minute film *The Holy Geometry* for general audiences; a new

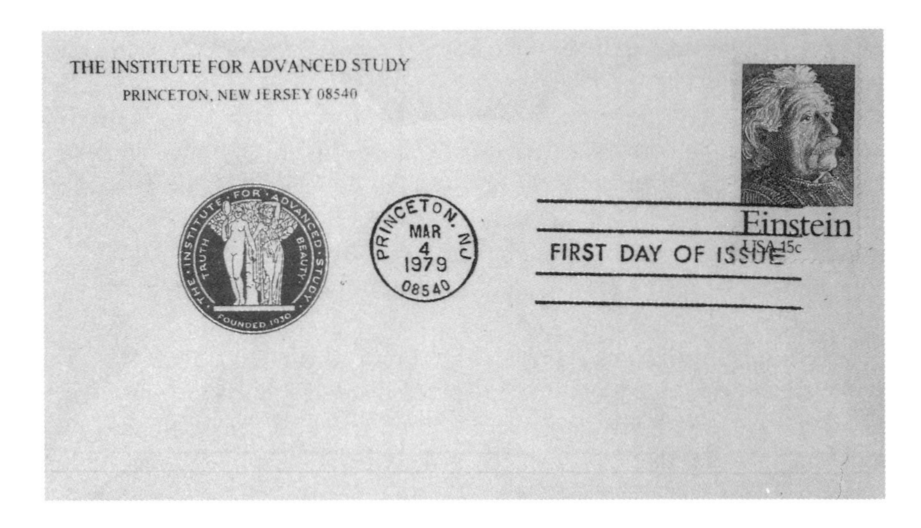

First day cover, used on the occasion of the Einstein centennial, March 14, 1979. (Institute for Advanced Study, Princeton.)

biography, *Albert Einstein, a partial portrait*, published by Scribner's Sons, New York; a traveling exhibit: 'Albert Einstein: image and impact', prepared by the American Institute of Physics; the Albert Einstein Centennial Lecture Bureau, with the support of the National Endowment of the Humanities; and the Einstein Centennial Exhibit in the National Museum of History and Technology, Washington, D.C. Those events led of course to outpourings in the press; these I have not followed.

All that national and international hullabaloo did not reveal much new about Einstein the man. From that point of view, the most interesting centennial contributions I know of are found in the special commemorative issue published on 14 March 1979 by *The Princeton Packet*, a local weekly. It contains numerous recollections about the great man by other Princeton residents. To conclude this entire essay I shall quote from their sayings.

From a resident who was a naval officer during the Second World War, a recollection from that time:

> My C.O. told me to request an interview for the purpose of obtaining Einstein's opinion as to whether or not a secretary, formerly employed by him and now serving in the Navy, was fully qualified to be entrusted with confidential information.
>
> The interview was granted and held in Dr. Einstein's cluttered second floor study at the back of his Mercer Street residence. In response to my inquiry, he said, 'Yes I remember this girl as a loyal American who was conscientious, reliable and trustworthy in all re-

spects – but did you say she was wanted by the Naval Intelligence? I cannot recommend her for the Naval Intelligence.'

When I asked, 'Why not?', he said, 'Because she is not intelligent.'

From a woman who as a young girl used to go caroling at Christmastime, and once did so in front of Einstein's home:

> Presently the old gentleman came slowly down the icy steps to the sidewalk where we stood and, in the glow of the streetlamp, told us in a heavily-accented voice how much he had enjoyed our rendering of the old carols and asked one or two of us our names and about our school.
>
> It was all over in a few moments and the head of bushy white hair, wrapped in its long muffler, retreated to the warmth of the house. But the warmth within each of us was now equal to that of the fireside and we continued on our appointed route with a personal radiance.

From a resident who at that time had a baby daughter:

> Walking toward us were the two intellectual giants of this century – Albert Einstein and Bertrand Russell . . . As I wheeled another daughter about Princeton, Einstein would frequently peer inside, sometimes tickling her under the chin. On brisk, pre-winter days he still wore his Navy woolen cap, his long hair flowing out from beneath it.
>
> The son of a friend of mine was taught by Einstein how to ride a bicycle. What a rare demonstration of dynamic equilibrium by a leading authority in that branch of physics.

From his doctor:

> After the war, I limited my practice to Ophthalmology, and was happy to make certain that the Einstein family were able to see well. The Professor insisted on a regular appointment, and sat serenely in the waiting room awaiting his turn. He wanted no special favors.
>
> When I finished my yearly check-up of his eyes, I would tell him that his glasses could be improved 40 percent with a new prescription. His reply invariably, with a smile, would be, 'A friend in New York sends me these simple magnifying glasses as a gift each year, and if they do no real harm, Henry, I prefer not to change them. I don't want to hurt his feelings. You don't mind if I continue, do you?'
>
> So, year after year, I would have the pleasure of checking the eyes of this great stubborn man, knowing that my final prescription was of academic interest only.

From a woman who owned a flower shop:

> When he would pay his bill with his check I would save them. I thought the autograph was worth more than the check. When I had accumulated quite a few, Dr. Einstein telephoned and asked if I would cash the checks, so he could balance his check book.
>
> He also offered to provide us with as many of his autographs as I wished.

Finally, from his resident photographer:

> I always found him very gracious and patient. But he did not like to have his picture taken. He never wanted personal aggrandizement.
>
> One day when it was particularly important that I get a shot of him, he didn't want me to take it. He asked: 'What do you want my picture for? They know what I look like.'
>
> 'Have you ever stopped to think that a photograph can be very kind?' he asked one day. 'I don't. How do you mean kind?' 'A photograph never gets old', mused Dr. Einstein. 'You and I change, people all change through the months and years, but a photograph always remains the same. How nice to look at a photograph of mother and father taken many years ago. You see them as you remember them.'

And the show goes on. . . .

References

Note
The following abbreviations are used:
NYT, *New York Times*.
SL, A. Pais, *Subtle is the Lord*, Clarendon Press, Oxford, 1982.
VZ, *Vossische Zeitung*.

1. Cf. O. Glasser, *W. C. Roentgen*, 2nd edn, Springer, Berlin, 1959; English translation by C.C. Thomas, Springfield, Ill. 1934.
2. P. Frank, *Einstein, sein Leben und seine Zeit*, p. 290, Vieweg, Braunschweig, 1979.
3. A. Einstein, *Ideas and opinions*, p. 15, Crown Publishers, New York, 1982.
4. *Anzeiger für die Stadt Bern*, 5 February 1902, Section 3, p. 2; repr. in *The collected papers of Albert Einstein*, Vol. 1, p. 334, ed. J. Stachel *et al.*, Princeton University Press, Princeton, 1987.
5. *The comparative reception of relativity*, ed. T.F. Glick, Reidel, Boston, 1987.
6. See *SL*, p. 150.
7. *Neue Zürcher Zeitung*, 10 March 1909.
8. *Berner Tageblatt*, 10 May 1909.
9. Cf. *SL*, Chapter 30.
10. *Prager Tagblatt*, 15 January 1911; *Bohemia*, same date.
11. *Prager Tagblatt*, 22 January 1911.
12. *Prager Tagblatt*, 23, 24 May 1911; *Bohemia*, 24 May 1911.
13. *SL*, Chapter 11.
14. *Der Bund*, 31 January 1912.
15. *Frankfurter Zeitung*, 3 February 1912.
16. *Prager Tagblatt*, 30 July 1912.

17. *Neue Freie Presse*, 5 August 1912; excerpted in *Prager Tagblatt*, same date.
18. C. Kirsten and H. J. Treder, *Albert Einstein in Berlin, 1913–1933*, Vol. I, p. 95.
19. Ref. 18, p. 98.
20. A. Einstein, letter to J. Laub, 22 July 1913, Einstein Archive.
21. *Vossische Zeitung* (referred to hereafter as *VZ*), 1 August 1913.
22. *VZ*, 2 January 1914.
23. Cf. W.L. Shirer, *The rise and fall of the Third Reich*, p. 245, Simon and Schuster, New York, 1960.
24. *VZ*, 26 April 1914.
25. *VZ*, 23 May 1917.
26. *VZ*, 25 June 1914.
27. *VZ*, 3 July 1914.
28. *VZ*, 6 November 1914.
29. *VZ*, 27 July 1914.
30. *VZ*, 20 August 1914.
31. *VZ*, 18 February 1913.
32. *SL*, Chapter 16, Section (c). See also J. Grelinsten, *The Physics Teacher*, **18**, 115, 187, 1980.
33. *SL*, p. 200.
34. *Nieuwe Rotterdamsche Courant*, 9 November 1919. See also *ibid.* 11 and 19 November.
35. Max Born in the *Frankfurter Allgemeine Zeitung*, 23 November 1919.
36. Cf. *SL*, pp. 309–10.
37. *Nieuwe Rotterdamsche Courant*, 4 July 1921.
38. *NYT*, 28 January 1928.
39. For their beginnings, see *SL*, Chapter 16, Section (d).
40. *NYT*, 2 February 1920.
41. *NYT*, 6 March 1927, Section VIII.
42. *NYT*, 9 April 1929.
43. *NYT*, 18 August 1929, Section V.
44. Also in translation in *Neue Zürcher Zeitung*, 8 January 1920.
45. See *SL*, p. 504.
46. See e.g. the *Arbeiter Zeitung*, Vienna, 15 December 1920.
47. A. Einstein, letter to P. Ehrenfest, 4 December 1919.
48. See further *SL*, Chapter 16, Section (d).
49. *Vorwärts*, 13 February 1920.
50. *Vorwärts*, 14 February 1920.
51. *100 Autoren gegen Einstein*, ed. H. Israel *et al.*, Voigtlander, Leipzig, 1931.
52. *Berliner Tageblatt*, 27 August 1920.
53. M. von Laue, W. Nernst, and H. Rubens, *Tägliche Umschau*, 26 August 1920.

54. A. Einstein, letter to M. Born, 9 September 1920.

55. E.g. in *Prager Tagblatt*, 29 August 1920; *Morning Post*, London, 2 September 1920.

56. *NYT*, 2 April 1921.

57. *NYT*, 3 April 1921.

58. *NYT*, 9 April 1921.

59. *NYT*, 11 April 1921.

60. *NYT*, 16 April 1921.

61. *NYT*, 26 April 19221.

62. *Berliner Tageblatt*, 19 July 1921.

63. *NYT*, 27 April 1921.

64. *NYT*, 10 May 1921; also *Daily News*, same day.

65. *NYT*, 18 May 1921.

66. *NYT*, 19 May 1921.

67. *NYT*, 26 May 1921.

68. *NYT*, 31 May 1921.

69. *NYT*, 9 June 1921.

70. *The Times* (London), 14 June 1921.

71. *NYT*, 2 July 1921.

72. *NYT*, 8 July 1921.

73. *NYT*, 12 July 1921.

74. *Berliner Tageblatt*, July 1921, date illegible on my copy.

75. *VZ*, 10 July 1921.

76. *Petit Journal*, 29 March 1922.

77. *Le Matin*, 29 March 1922.

78. *Echo National*, 30 March 1922.

79. *Berliner Tageblatt*, 12 April 1922.

79*a*. C. Harnist, letter to A. Einstein, 31 March 1929, Einstein Archive.

80. *Die Umschau*, Frankfurt am Main, 16 April 1922.

81. *L'Oeuvre*, early April 1922, date illegible on my copy; quoted in *NYT*, 5 April 1922.

82. *NYT*, 6 April 1922.

83. *Echo de Paris*, 3 April 1922.

84. *Le petit Parisien*, 10 April 1922.

84*a*. *NYT*, 12 June 1922.

85. *Neue Rundschau* **33**, 815, 1922.

86. Ref. 3, p. 187.

87. *The New Republic*, **32**, 197, 1922.

88. *Berliner Tageblatt*, 5 August 1922.

89. Ref. 18, Vol. I, p. 231.

90. *NYT*, 16 November 1922.

91. *Singapore Daily*, 3 November 1922.

92. *The Eastern Times*, 10 November 1922.

93. *Ibid.*, 11 November 1922.

94. Ref. 18, document 153.

95. Ref. 5, p. 351.

96. J. Ishiwara, *Einstein Kōen-Roku*, Tokyo-Tosho, Tokyo, 1977.

97. T. Ogawa, *Jap. St. Hist. Sci.*, **18**, 73, 1979.

98. *SL*, Chapter 30.

99. A. Pais, *Niels Bohr's times, in physics, philosophy, and polity*, Chapter 10, Section (f), Clarendon Press, Oxford, 1991.

100. *The Palestine Weekly*, 9 February 1923.

101. *SL*, p. 38.

102. *NYT*, 30 March 1927.

103. C. Alpert, letter to the *Jerusalem Post*, 10 May 1990.

104. Cf. *SL*, pp. 35, 36.

105. A. Einstein, letter to M. Grossmann, 3 January 1908, Einstein Archive.

106. C. Stoll, letter to H. Ernst, 4 March 1909, Einstein Archive.

107. A. Einstein, *On Zionism*, pp. 41, 43, transl. L. Simon, Mcmillan, New York, 1931.

108. K. Blumenfeld, letter to C. Weizmann, 15 March 1921; *ETH Bibl. Zürich Hs* **304**, 201–4.

109. A. Einstein, letter to K. Blumenfeld, 25 March 1955.

110. *The New Palestine*, 27 March 1925, reprod. in ref. 3, p. 63.

111. *Jüdische Rundschau*, **30**, 129, 1925.

112. *NYT*, September 25, 1924.

113. T.F. Glick, *Einstein in Spain*, Princeton University Press, 1988.

114. Ref. 113, p. 327.

115. Ref. 113, pp. 325–6.

116. *NYT*, 20 March 1923.

117. Ref. 18, document 154.

118. See ref. 113, pp. 357–74; also ref. 5, p. 231.

119. Ref. 113, p. 148.

120. *NYT*, 25 March 1925.

121. Ref. 18, document 156.

122. Ref. 5, p. 381.

123. A. Einstein, *Revista matematica Hispano-Americana*, **1**, 72, 1926.

124. *NYT*, 5 June 1925.

125. O. Nathan and H. Norden, *Einstein on peace*, Schocken, New York, 1968.

126. F. Gilbert, *The end of the European era, 1890 to the present*, 2nd edn, p. 137, Norton, New York, 1979.

127. *Berliner Tageblatt*, 17 October 1919, reprod. in H. Wehberg, *Wider den Aufruf der 93*, p. 31, Deutsche Verlagsges. für Politik und Geschichte, Berlin, 1920.

128. A. Einstein, letter to H.A. Lorentz, 1 August 1919, reprod. in ref. 125, p. 33.

129. C. Seelig, *Albert Einstein*, p. 15, Europa Verlag Zürich, 1960.

130. Ref. 125, p. 7.

131. For the full text see ref. 125, pp. 4–6.

132. Ref. 125, p. 74.

133. Wehberg, ref. 127, p. 22 ff.

134. For more on the *Bund*, see ref. 125, pp. 9–12.

135. Ref. 18, Vol. 1, doc. 118.

136. Ref. 125, p. 17.

137. Ref. 18, Vol. 1, docs. 59–68.

138. Ref. 18, Vol. 1, docs. 81–87.

139. A. Einstein, letter to H. Zangger, undated, probably spring 1915.

140. Ref. 125, p. 25.

141. Ref. 125, pp. 32, 36.

142. Ref. 125, pp. 30, 31.

143. *New York Evening Post*, 26 March 1921.

144. *Christian Century*, July 1929.

145. *Berliner Tageblatt*, 16 May, 9 June, 1922; *VZ*, 14 June 1922.

146. *NYT*, 28 June 1923; also *VZ*, 22 March 1923.

147. A. Einstein, letter to G. Murray, 30 May 1924, reprod. in ref. 125, p. 66.

148. *VZ*, 28 July 1924.

149. *Frankfurter Allgemeine Zeitung*, 29 August 1924.

150. *Pressedienst der Deutschen Liga für Völkerbund*, 10 December 1926.

151. A. Reiser, *Albert Einstein*, Boni, New York, 1930.

152. Ref. 125, p. 111.

153. *New York Evening Post*, 1 January 1926.

154. *NYT*, 17 May, 1925.

155. Ref. 18, Vol. 1, doc. 136.

156. Ref. 18, Vol. 1, doc. 140.

157. A. Einstein, statement prepared for the *Liga der Menschenrechte*, 6 January 1929.

158. A. Einstein, letter to K.R. Leitner, 8 September 1932.

159. A. Einstein, letter to V. Molotov, 23 March 1936.

160. A. Einstein, letter to V. Molotov, 4 July 1936.

161. A. Einstein, letter to J. Stalin, 17 November 1947.

162. S.K. Tsarapkin, letter to A. Einstein, 18 December 1947.

163. Cf. *SL*, Chapter 13.

164. For references to these and other similar writings see *SL*, Chapter 16, Section (e).

165. H. Frentz, in *Die Furche* (Vienna), 5 April 1969.

166. *NYT*, 6 November 1927.

167. *NYT*, 3 May 1928.

168. *NYT*, 18 April 1926.

169. *NYT*, 6 March 1927, Section VIII.

170. A. Einstein, letter to J. Hadamard, 24 September 1929.

171. Cf. ref. 125, Chapter 4.

172. *NYT*, 21 January 1930.

173. *Die Menschenrechte*, 20 July 1930.

174. Statement submitted to the Danish paper *Politiken*, 5 August 1930.

175. *NYT*, 21 December 1930; also *Bund*, 17 December 1930; reprod. in its entirety in *Albert Einstein, Cosmic Religion*, p. 57, Covici-Friede, New York, 1931.

176. *NYT*, 22 November 1931.

177. *The Friend*, 12 August 1932.

178. *Jugendtribune*, 17 April 1931.

179. *NYT*, 28 February 1932.

180. *Berlin am Morgen*, 13 October 1932.

181. *Die Menschenrechte*, 20 March 1931.

182. *NYT*, 17 February 1931.

183. *NYT*, 2 August 1931.

184. *The Nation*, 23 September 1931.

185. *NYT*, 26 January 1932.

186. *Die Menschenrechte*, 20 August 1931.

187. *NYT*, 5 March 1931.

188. *The New World*, July 1931 issue.

189. A. Einstein, *Sitz. Ber. Preuss. Ak. Wiss*. 1929, p. 2.

190. *NYT*, 12 January 1929.

191. A.S. Eddington, letter to A. Einstein, 11 February 1929.

192. *NYT*, 4 February 1929.

193. *NYT*, 13 March 1929.

194. *NYT*, 21 April 1929.

195. Details about the Caputh affair, including the text of Einstein's two letters are in *Der Tagesspiegel*, 20 April 1955.

196. *NYT*, 17 June 1930.

197. *NYT*, 23 August 1930.

198. *NYT*, 14 September 1930.

199. *News Chronicle*, 28 October 1930.

200. Reproduced in full in *Cosmic Religion* (ref. 175), p. 84.

201. *NYT*, 29 October 1930. For the full text of Shaw's toast, see e.g. *Berliner Tageblatt*, same date.

202. *Natal Mercury* (Durban), 4 November 1930.

203. *NYT*, 3 December 1930.

204. *NYT*, 12 December 1930.

205. See e.g. *NYT*, 21 December 1930.

206. *NYT*, 15 December 1930.

207. *NYT*, 16 December 1930.

208. *Liberty Magazine*, 9 January 1932.

209. *NYT*, 1 January 1931.

210. *The Engineer*, October 1979.

211. *NYT*, 15 March 1931.

212. *NYT*, 17 April 1931.

213. *NYT*, 5 October 1931.

214. *Pictorial Review*, February 1933.

215. Ph. Frank, *Einstein*, p. 361, Paul List Verlag, Munich, 1949.

216. A. Einstein, in his personal travel diary, 6 December 1931.

217. Ref. 18, Vol. 1, doc. 146.

218. *New York World Telegram*, 11 March 1933.

219. *NYT*, 19 April 1955.

220. The Institute for Advanced Study, excerpt from minutes, 10 October 1932.

221. My edition was published by Diogenes Verlag, Zürich 1972.

222. *SL*, p. 514.

223. A. Einstein, letter to A. Bachrach, 25 July 1949.

224. Ref. 18, Vol. 1, doc. 161.

225. *Aufbau*, 9 March 1979.

226. Ref. 18, Vol. 1, doc. 163.

227. *NYT*, 16 March 1933.

228. Ref. 18, Vol. 1, doc. 169.

229. E.g. *Neue Zeit*, 2 April 1933; also ref. 18, Vol. 1, doc. 173.

230. Ref. 18, Vol. 1, doc. 181; also *NYT*, 12 April 1933.

231. *Journal Officiel*, session of 13 April 1933, p. 2276; also *Arbeiter Zeitung*, 14 April 1933. See further letters by the French Consul in Ostende to Einstein, 8 and 13 April 1933, Einstein Archive.

232. *Le Monde*, 22 April 1955.

233. Ref. 113, Chapter 9.

234. *Bunte Woche* (Vienna), 1 October 1933.

235. *New York World Telegram*, 19 September 1933.

236. *NYT*, 10 September 1933.
237. *NYT*, 4 October 1933.
238. A. Flexner, letter to A. Einstein, 13 October 1933.
239. A. Einstein, *Ideas and Opinions*, p. 7, Crown Publishers, New York, 1954.
240. *NYT*, 13 October 1937.
241. *NYT*, 27 July 1973.
242. A. Einstein, letter to Queen Elizabeth, 20 November 1933.
243. A. Einstein, letter to Queen Elizabeth, 16 February 1935.
244. *NYT*, 23 June 1940; in more detail in ref. 125, pp. 312–14.
245. *New York Sun*, 9 January 1935. The full text is in ref. 125, p. 254.
246. *NYT*, 19 December 1933.
247. *NYT*, 18 March 1934, Section 4.
248. *NYT*, 14 April 1934, Section 2.
249. *NYT*, 30 April 1934.
250. *Pittsburgh Post–Gazette*, 29 December 1934; *The Literary Digest*, 12 January 1935.
251. *NYT*, 13 April 1935.
252. *NYT*, 15 April 1935.
253. A. Einstein, *Out of my later years*, p. 269, The Citadel Press, Secaucus, New Jersey, 1977.
254. *NYT*, 27 April 1935.
255. *NYT*, 29 June 1935.
256. *NYT*, 23 October 1935.
257. Ref. 125, p. 262.
258. For a detailed biography of Ossietzky, see K.R. Grossmann, *Ossietzky*, Kindler Verlag, Munich, 1963.
259. *New York World Telegram*, 31 May 1935; *New Yorker Volkszeitung*, 15 June 1935; *New York Post*, 24 June 1935.
260. Ref. 253, p. 241.
261. *NYT*, 12 February 1936.
262. *NYT*, 28 March 1936.
263. *NYT*, 9 March 1936.
264. *NYT*, 16 October 1936.
265. For the full text, see ref. 239, p. 59.
266. Ref. 253, p. 5.
267. *NYT*, 19 April 1937; also ref. 125, p. 274.
268. *The Brooklyn Tablet*, 14 May 1938.
269. *NYT*, 7 June 1938. The complete text of the address is found in ref. 253, p. 15.
270. *NYT*, 16 September 1938.
271. See e.g., ref. 239, p. 18; ref. 253, p. 11.

272. *NYT*, 30 October 1938.

273. *NYT*, 29 January 1939.

274. For an English translation of that paper, see H.G. Graetzer, *Am. J. Phys.* **32**, 9, 1964.

275. L. Meitner and O.R. Frisch, *Nature*, **143**, 239, 1939.

276. *NYT*, 14 March 1939.

277. *NYT*, 1 May 1939.

278. *NYT*, 29 May 1939.

279. *NYT*, 2 July 1939.

280. *NYT*, 17 August 1939.

281. *NYT*, 11 November 1939.

282. R. Rhodes, *The making of the atomic bomb*, p. 305, Simon and Schuster, New York, 1986.

283. This letter has often been reproduced in full; see e.g. ref. 125, p. 295.

284. Ref. 125, p. 297.

285. Ref. 125, p. 299.

286. A. Pais, *Niels Bohr's Times*, p. 482 ff., Clarendon Press, 1991.

287. H.L. Stimson, *On active duty in peace and war*, Chapter 13, Harper, New York, 1947.

288. A. Vallentin, *The drama of Albert Einstein*, p. 278, Doubleday, New York, 1954.

289. A. Einstein, letter to N. Bohr, 12 December 1944.

290. *NYT*, 19 March 1940.

291. *NYT*, 23 May 1940.

291*a*. NYT, 23 June 1940

292. *NYT*, 16 June 1941.

293. Ref. 125, p. 320.

294. *NYT*, 30 December 1941.

295. *NYT*, 26 October 1942.

296. *NYT*, 12 March 1944.

297. *NYT*, 29 May 1944.

298. *NYT*, 5 August 1944.

299. *NYT*, 10 October 1944. For the full text of Einstein's statement, see ref. 125, p. 332.

300. *Aufbau* [a New York publication], 27 April 1945.

301. *NYT*, 12 August 1945.

302. *NYT*, 1 January 1949.

303. *NYT*, 13 January 1949.

304. *NYT*, 14 January 1949.

305. *NYT*, 15 March 1949.

306. *NYT*, 14 March 1950.

307. *NYT*, 14 March 1949.

308. *NYT*, 16 March 1949.

309. Ref. 253, p. 272. See also ref. 317.

310. *NYT*, 20 March 1949.

311. *NYT*, 17 March 1953.

312. *NYT*, 16 March 1953.

313. *NYT*, 14 March 1954.

314. *NYT*, 15 March 1954.

315. *NYT*, 27 December 1949.

316. *NYT*, 28 December 1949.

317. *NYT*, 2 January 1950.

318. *NYT*, 8 January 1950.

319. *NYT*, 9 January 1950.

320. *NYT*, 15 February 1950.

321. *NYT*, 11 May 1950.

322. *NYT*, 30 March 1953.

323. *NYT*, 19 November 1952.

324. Y. Navon, in *Albert Einstein, Historical and Cultural Perspectives*, p. 293, Princeton University Press, 1982.

325. D. Mitrany, in *Jewish Observer and Middle East Review*, 22 April 1955. For other details, see ref. 125, pp. 571–4.

326. *NYT*, 4 May 1946.

327. *NYT*, 28 April 1948.

328. Full text in ref. 239, p. 146.

329. *NYT*, 5 May 1953.

330. Full text in ref. 239, p. 33.

331. *NYT*, 21 February 1954.

332. *NYT*, 6 November 1949.

333. *NYT*, 14 May 1951.

334. *NYT*, 23 September 1946.

335. *NYT*, 25 September 1946.

335*a*. *NYT*, 16 October 1947.

336. *NYT*, 30 March 1948.

337. *NYT*, 13 March 1949.

338. *NYT*, 11 August 1949.

339. *NYT*, 12 March 1950.

340. *NYT*, 6 January 1951.

341. *NYT*, 15 January 1951.

342. *NYT*, 24 February 1951.

343. *NYT*, 15 March 1951.
344. *NYT*, 28 August 1951.
345. *NYT*, 16 March 1952.
346. *NYT*, 5 April 1952.
347. *NYT*, 16 May 1952.
348. *NYT*, 17 May 1952.
349. *NYT*, 10 August 1952.
350. *NYT*, 5 October 1952.
351. *NYT*, 26 October 1952.
352. *NYT*, 29 November 1952.
353. *NYT*, 17 January 1953.
353*a*. *NYT*, 3 July 1954.
354. *NYT*, 25 May 1953.
355. *NYT*, 11 October 1953.
356. *NYT*, 24 December 1953.
357. *NYT*, 15 September 1945.
358. *NYT*, 10 October 1945; full text in ref. 125, p. 340.
359. *Congressional Record of 1945*, Vol. 91, part 8, p. 10049.
360. *Atlantic Monthly*, November 1945. Full text in ref. 239, p. 118.
361. *NYT*, 11 December 1945.
362. Full text in ref. 239, p. 115.
363. *NYT*, 24 February 1946.
364. *NYT*, 30 May 1946.
365. *NYT*, 23 June 1946. Full text in ref. 125, p. 383.
366. *NYT*, 18 November 1946.
367. *NYT*, 19 August 1946.
367*a*. *NYT*, 22 February 1947. The message was broadcast on station WMCA. Its full text is in ref. 125, p. 403.
368. *NYT*, 16 July 1947.
369. 'The military mentality', in *The American Scholar*, New York, summer 1947; full text in ref. 239, p. 132.
370. *NYT*, 23 September 1947.
371. Full text in ref. 253, p. 156.
372. *Atlantic Monthly*, November 1947. Full text in ref. 239, p. 123.
373. *NYT*, 29 August 1948.
374. *NYT*, 13 February 1950.
375. *The Evening Bulletin*, 14 February 1950.
376. *Newark Evening News*, 14 February 1950.
377. *NYT*, 13 October 1952.
378. *NYT*, 15 October 1952.

379. *NYT*, 13 January 1953.

380. *NYT*, 12 June 1953.

381. See e.g. *NYT*, 17 June (*pro*) and 18 June (*con*) 1953.

382. *NYT*, 14 June 1953.

383. *NYT*, 22 June 1953.

384. Ref. 382, Sec. IV, p. 2.

385. *NYT*, 26 June 1953.

386. *NYT*, 19 and 22 December 1953.

387. *The Washington Post*, 13 June 1953.

388. *The Detroit News*, 15 June 1953.

389. *NYT*, 14 March 1954.

390. *NYT*, 11 April 1954.

391. *NYT*, 20 August 1954.

392. *The Reporter*, 18 November 1954.

392*a*. *Newark Star Ledger*, 14 April 1954.

393. *Bull. At. Scientists*, May 1954.

394. *The New Palestine*, **11**, 334, 1926.

395. Letter to the *Manchester Guardian*, 12 October 1929; full text in ref. 107, p. 71.

396. Letter to the Palestinian Arab paper *Falastin*, 28 January 1930; full text in ref. 107, p. 87.

397. *NYT*, 28 September 1930.

398. Full text in ref. 239, p. 174.

399. *NYT*, 21 April, Section 2, 1954.

400. Reproduced in ref. 253, p. 257. I do not know where this statement was published originally.

401. *Aufbau*, 16 March 1979.

402. *NYT*, 21 April 1935.

403. *NYT*, 8 June 1936.

404. *NYT*, 28 January 1938.

405. *NYT*, 18 April 1938. Full text in ref. 239, p. 188.

406. *Jerusalem Post*, 11 November 1988.

407. *Collier's Magazine*, 26 November 1938. Full text in ref. 239, p. 191.

408. Full text in ref. 239, p. 198.

409. *NYT*, 28 May 1939.

410. *Technion Journal*, April 1941.

411. *NYT*, 6 June 1944.

412. *NYT*, 12 January 1946.

413. *NYT*, 15 February 1946.

414. *NYT*, 26 November 1947.

415. *NYT*, 18 April 1948, letter to the Editor, co-signed with Leo Baeck.

416. *NYT*, 16 March 1949.

417. *NYT*, 18 November 1949.

418. *NYT*, 28 November 1949. For the full text of the broadcast, see ref. 239, p. 200.

419. *NYT*, 30 October 1951.

420. *NYT*, 25 May 1953.

421. *NYT*, 20 September 1954.

422. Ref. 125, p. 639.

423. Ref. 239, p. 171.

424. Ref. 239, p. 186.

425. *NYT*, 19 April 1955.

426. *NYT*, 21 April 1955.

427. *NYT*, 22 April 1955.

428. *NYT*, 15 May 1955.

429. *NYT*, 31 August 1955.

430. *News of the YIVO*, September 1979, No. 150.

431. *L'express*, June 7–13, 1965.

432. *The Jerusalem Post Special*, July 1989.

433. *Jerusalem Post*, 18 January 1980.

434. *Vegetarisches Universum*, December 1957.

435. A. Friedman and C. Donley, *Einstein as myth and muse*, Cambridge University Press, 1985.

436. Examples referring to Einstein's life are found in ref. 435.

437. L. Durrell, *Balthazar*, p. **5**, Faber and Faber, London, and E.P. Dutton, New York, 1958.

438. Ref. 437, p. 142.

439. L. Durrell, *Clea*, p. 135, Faber and Faber, London, and E.P. Dutton, New York, 1960.

440. English translations by J. Kirkup, Grove Press, New York, 1964.

441. All reproduced in *Einstein, a centenary volume*, p. 242, ed. A.P. French, Heinemann, London, 1979.

442. Repr. in ref. 435, p. 183.

443. *Neue Zürcher Zeitung*, 26 February 1979.

ONOMASTICON

Note: An asterisk(*) after a name indicates that it receives fuller treatment in the Index of Subjects.

SUBJECT INDEX